MONOGRAPHS ON SCIENCE, TECHNOLOGY, AND SOCIETY

MONOGRAPHS ON SCIENCE, TECHNOLOGY, AND SOCIETY

1. Eric Ashby and Mary Anderson: The politics of clean air
2. Edward Pochin: Nuclear radiation: risks and benefits
3. L. Rotherham: Research and innovation: a record of the Wolfson Technological Projects Scheme 1968–1981; with a foreword and postscript by Lord Zuckerman
4. John Sheail: Pesticides and nature conservation 1950–1975

NUCLEAR RADIATION: RISKS AND BENEFITS

Edward Pochin

Consultant to the Director, National Radiological Protection Board

CLARENDON PRESS · OXFORD

Oxford University Press, Walton Street, Oxford OX2 6DP
London New York Toronto
Delhi Bombay Calcutta Madras Karachi
Kuala Lumpur Singapore Hong Kong Tokyo
Nairobi Dar es Salaam Cape Town
Melbourne Auckland
and associated companies in
Beirut Berlin Ibadan Mexico City Nicosia

Oxford is a trade mark of Oxford University Press

Published in the United States
by Oxford University Press, New York

First published 1983
First issued in paperback 1985

British Library Cataloguing in Publication Data
Pochin, Edward
Nuclear radiation: risks and benefits. –
(Monographs on science, technology and society)
1. Ionizing radiation - Physiological effect
I. Title II. Series
612'.014486 QP82.2.I53
ISBN 0-19-858329-X
ISBN 0-19-858337-0 (pbk)

Set by Hope Services, Abingdon
Printed by St Edmundsbury Press,
Bury St Edmunds, Suffolk

PREFACE

This book is written for the general reader who wishes to examine the nature, the uses, and the hazards of radiation, but who is not necessarily familiar with the physical and biological processes which underlie these uses and hazards. It deals with the different forms of so-called ionizing radiations to which people are exposed from cosmic radiation, from x rays, and from radioactive materials in the environment or incorporated in the body. In particular, attention is given to the amounts of radiation received from different sources, the kinds and amounts of harm that may result, and the available ways of minimizing any such harm.

I am grateful to many of my colleagues and friends for advice and suggestions and in particular to Miss Pamela Bryant, Dr Anthony Searle, and Dr Giovanni Silini for their comments on sections of the text; to Mrs Heather Reynolds and Miss Linda Butler for their great care in preparation of the typescript; and to Sally, Charles, and Susan Pochin for pointing out some of the more incomprehensible passages in my drafts, and for encouragement generally.

I am grateful also for permission to reproduce Figs. 4.1 and 4.2 by the National Radiological Protection Board; Fig. 7.1 by the Cambridge University Press (from *Radiation from radioactive substances*, by Rutherford, Chadwick, and Ellis); Fig. 7.2 from Dr Michael Marshall and the Atomic Energy Research Establishment, Harwell; Fig. 8.1 from Mr George Brecken and the Medical Research Council's Radiobiology Unit at Harwell; and Fig. 10.1 from Dr Norman Stott and the Atomic Energy Research Establishment, Harwell.

I am most grateful also to Mr John Millham for his help in preparing photographic reproductions and to Mrs Jean Fisher for her preparation of figures.

December 1982 E.E.P.

CONTENTS

1

THE FIRST FIFTY YEARS

The value that radiation might have in medicine was clearly understood by Röntgen at the time of his discovery of x rays in 1895. The harm that could be caused by radiation exposure of the skin was recognized a few months later, and already by 1902 a cancer had developed on the hand of a worker in a factory making x-ray tubes.

For fifty years, medical radiology remained the only significant source of man-made radiation exposure. The increasing variety of diagnostic applications, however, and particularly the use of x rays and radioactive materials in treatment, demonstrated the dangers of over-exposure of patients or staff. The need to secure the fullest medical benefit without causing any undue harm was largely responsible, therefore, for an increasingly detailed study during the 1920s of the effects of radiation on living organisms. It led also to the development in 1928 of internationally accepted recommendations on bases for protection against radiation, and methods of measuring amounts of radiation exposure.

There was thus already quite extensive knowledge of the ways in which radiation caused harm, and experience in practical radiological protection, at the time when, in the mid-1940s, the possible sources of medical, occupational, and environmental exposure started to increase so rapidly in variety and extent, and when whole populations were exposed to lethal amounts of radiation in Hiroshima and Nagasaki, and to small amounts world-wide from atmospheric nuclear weapon tests subsequently.

The present variety of ways in which people may be exposed to radiation shows the importance of assessing the amounts of each such exposure, and the types and likely frequencies of any resulting harmful effects. The extensive studies that have been made, by physical, radiobiological, and medical methods, of the effects of ionizing radiation now make it possible to estimate these risks with considerable confidence, and certainly on a more reliable basis than is available for most harmful agents in the working or general environment.

RÖNTGEN AND X RAYS

The discovery of x rays depended upon a simple observation. Wilhelm Röntgen was examining the effects of passing an electric current through a glass tube from which the air had been partially evacuated. His 'Crookes tube' was covered with thin black cardboard, and the glow caused in the tube by the passage of the current was not visible in the darkened laboratory. On 8 November 1895, however, there happened to be a sheet of light-sensitive barium platino-cyanide paper on the bench, and this paper was seen to glow every time the current was switched on.

Röntgen quickly found that this 'radiation' emitted from the tube also caused exposure of photographic plates and that, although it passed through cardboard that was opaque to light (and, he records, through a book of a thousand pages or two packs of whist cards), it did not pass as readily through metal or various other dense objects. In his report to the Würzburg Physico-Medical Society on 28 December, he demonstrated that when his hand was placed between the tube and a fluorescent screen, the image on the screen showed that the radiation passed through the soft tissue of the hand with little diminution, but that the bones of the hand were sharply outlined.

The medical potentialities of x rays were rapidly and widely recognized. Reprints of his paper were sent out to scientific colleagues four days after his report was given in Würzburg, a promptness which most investigators today would find enviable. Extensive reports appeared in the European and American press in early January 1896, and radiographs had been produced in many countries before the end of the month.

The scope of x-ray applications in medical diagnosis depended, and still depends, on differences in absorbtion of x rays in different materials. The amount of absorbtion varies with the 'effective atomic number' of the materials through which they pass. 'Light' chemical elements, such as the hydrogen, carbon, nitrogen, and oxygen of which the soft tissues of the body are largely composed, have low atomic numbers, and materials of such composition transmit x rays relatively freely. Absorbtion is high, however, in materials composed of heavier chemical elements such as calcium, barium, iodine, and various metals – the degree of absorbtion in each case varying with the thickness of the material through which the x rays pass or the concentration of the heavier elements in it.

Accurate x-ray pictures of bone structure or of fractures could therefore be obtained as a result of the greater opacity to x rays of the calcium-rich bone than of the soft tissues. Useful information was also obtainable when the calcium concentration in bone was abnormal, either locally or generally. Thus local areas of translucency to x ray may result from sites of infection or tumour tissue, or when calcium is lost from the bones following decreased activity in a limb, or of the skeleton as a whole in certain diseases or in old age. On the other hand, increased opacity is seen during healing of fractures, and certain cancers of bone may reveal their presence by areas of dense new bone formation.

The reduced transmission of x rays through calcified material was also being used already by 1905 for the detection of calcium-containing stones in the gall-bladder or urinary tract, and a detectable calcification of tuberculous foci was recognized as indicative of healing.

Similarly, the reduced transmission of x rays through many metals was soon being used to detect and localize metallic fragments, including bullets, in body tissues; and the radiographs of Röntgen's wife's hand, preserved in various science museums, show clearly the structure of the ring which she wore.

The deliberate administration of materials containing the heavier chemical elements was used first to outline the stomach or intestine. A substance was swallowed which was opaque to x rays and was not absorbed from the gut. X rays were then taken at appropriate stages of its passage down the gastro-intestinal tract. After initial attempts with various rather toxic salts of bismuth and barium, the present conventional barium sulphate 'meal' was in routine use by 1931.

The uses of such 'contrast media' to show the outlines of the urinary passages from kidney to bladder, or of the gall-bladder and biliary ducts, came later. Such uses depended on finding some substance which would be excreted in concentrated form by the kidney or gall-bladder into these passages, which would not be toxic, and which would contain a chemical element of high enough atomic mass to outline the ureters or biliary ducts during its passage through them. Various iodine-containing compounds were developed and tested for these purposes, and were used also for outlining the bladder, the bronchial passages, and ultimately the cerebro-spinal fluid canal after direct injection into these various cavities.

One radiopaque material which gave good indication of any interruption or distortion of the blood-vessels in x rays taken during its injection into an artery was a suspension of thorium oxide known as Thorotrast. Thorium, being an element of high atomic mass, gave good radiographs of the arterial distribution. Unfortunately, its high atomic mass meant also that it was radioactive, and the harmful effects of its use, which has long been abandoned, are discussed later (p. 140).

One of the greatest values of x rays in diagnostic medicine depended, however, not on the difference in transmission between soft tissues and bone, metal, or injected materials of high atomic mass, but on the differences in transmission between soft tissues and air. Particularly in the days when lung infections were common and often fatal, the use of x rays allowed a diagnosis and a localization of disease which had not been achieved by the stethoscope or the percussing finger. Chest x rays could now detect the contrast between normal air-filled lung and the denser areas of tuberculous infection at the lung apex, of pneumonia at the lung base, or of an effusion from pleurisy surrounding and collapsing the lung. Moreover, the organs which could be silhouetted against the surrounding lung were themselves of the greatest medical importance: the heart, and its blood-vessels passing to the lung or to the head and arms; the oesophagus in its passage through the chest; and the diaphragm between the lungs and the abdomen, whose shape may be distorted by abnormalities within the abdomen. Even an 'aneurismal' weakening of the wall of the aorta, in the part of the artery lying within the chest, was detected by x ray in 1897, an impressive achievement for the second year of the use of x rays.

The properties of x rays in passing readily through some materials but being stopped by others, in causing fluorescent materials to glow, and in fogging photographic plates, suggested analogies with the behaviour of visible light, and with ultraviolet and infra-red radiations. It was indeed established that x rays also were due to electromagnetic waves, differing essentially from those of visible light merely by having a much shorter wavelength and a correspondingly higher energy. There is in fact a continuous range of these electromagnetic radiations, from gamma radiation (p. 9) and x rays, of short wavelength, through the ultraviolet, the whole spectrum of visible light, and the infra-red, and on up to radio waves of long wavelength (Fig. 1.1), our eyes being sensitive only to those with wavelengths in the centre of this range.

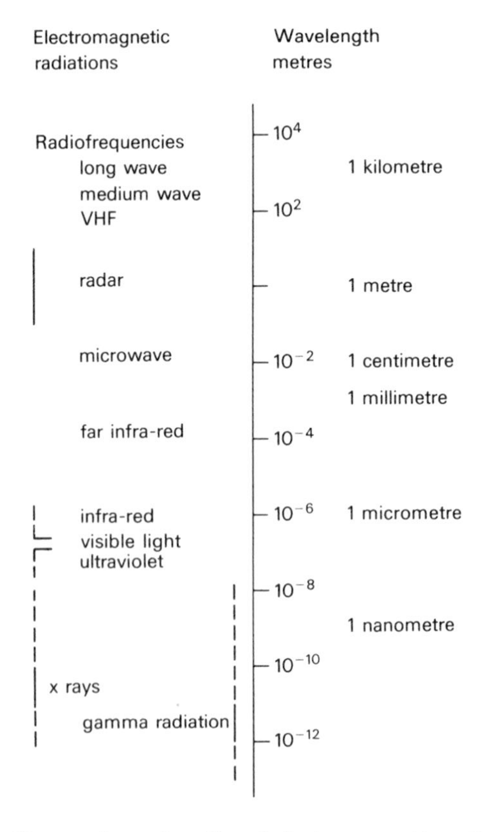

Fig. 1.1. Range of wavelengths of electromagnetic radiations.

In keeping with their higher energy, x rays have the effect, which radiations in the visible, ultraviolet, or infra-red ranges do not, of causing 'ionization' in the materials through which they pass. Their energy is sufficient to knock electrons from the atoms in these materials, leaving the atomic residues in the form of charged 'ions'. These free ions are instrumental in causing characteristic kinds of chemical damage, particularly in the chromosomes of the cells of body tissues. The resultant forms of harm are discussed in detail later (in Chapters 7, 8, and 9).

BECQUEREL AND RADIOACTIVITY

Other types of radiation, which proved to have a similar ionizing effect, were identified within a few months of the discovery of x rays. In February 1896, Henri Becquerel was examining various substances which were known to fluoresce after exposure to sunlight, to see whether x rays were emitted while they were fluorescing in this way. No such effect was found, but it was discovered that one of the substances tested, a compound of uranium, did cause darkening of the photographic plate. Moreover, it did so whether it had been exposed to sunlight or not.

Becquerel's discovery of radioactivity in uranium was rapidly followed by the identification by Pierre and Marie Curie of other chemical elements which spontaneously emitted radiation in this way; radium, thorium, and polonium were isolated chemically and found to have these properties in 1898.

RUTHERFORD AND THE ATOMIC NUCLEUS

To understand the nature of these radioactive materials and the kinds of radiation which they emit, we need to digress at this point to describe the understanding of atomic structure which developed during the first decades of the present century.

Atoms had long been thought of as the indivisible units of which all matter was composed. The atom was known to consist of a heavy, and positively charged, nucleus with a number of light and negatively charged electrons orbiting round it. The charge on the electrons equalled that on the nucleus so that the atom as a whole was uncharged. In some circumstances, one or more electrons might be temporarily stripped off, leaving the remainder of the atom in a charged, or ionized, state. The nucleus, however, remained intact.

Then in 1918, Ernest Rutherford made the first of a series of observations which was to show what the nucleus was made of. He had been exposing different gases to the radiation from a radioactive source, to see what kinds of particle emerged from the bombardment. With hydrogen, he had found no sign of any particle that could not have resulted from the hydrogen itself, or from the bombarding (alpha) radiation. With air, however, he found evidence of a new particle released which could not be due either to the incident radiation or to any of the atoms present in the air. Detailed analysis of his findings, and of the 'vapour trails' formed by the particles as they passed through the moist air of a 'cloud chamber', showed the new particle to be a component – a proton – knocked out of the nucleus of atoms of nitrogen in the air.

Rutherford's work with Soddy and others, and that of J.J. Thomson and Aston, were to show the components of which the nuclei of different chemical elements were made up, and the changes in nuclear composition that took place during radioactive decay.

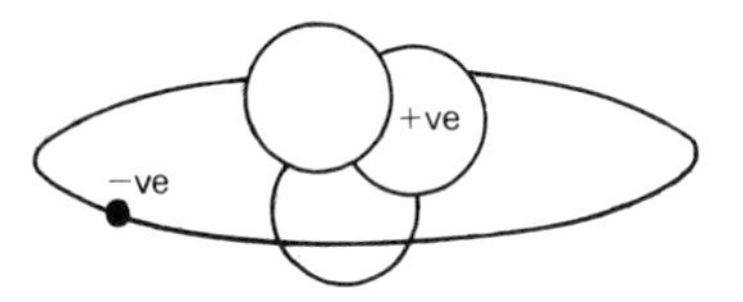

Fig. 1.2. Particles constituting the atom of tritium (hydrogen-3): one proton and two neutrons in the nucleus; one electron in orbit round the nucleus.

The nature of radioactivity

Radioactivity results from instability of the atomic nucleus. In all atoms, the nucleus is made up of charged and uncharged particles that are of about equal mass, the protons and neutrons respectively (Fig. 1.2). The chemical behaviour of an atom depends essentially upon the number of the charged protons in its nucleus, since this determines the size of the electrical charge on the nucleus and hence the number of electrons held in orbit around it. The number of protons, or 'atomic number' of a chemical element, ranges from one for hydrogen, with only one proton in the nucleus, to over one hundred for various highly unstable elements (Table 1.1).

Table 1.1

Examples of chemical elements, with their atomic numbers and their range of mass numbers (for all isotopes which are stable or with half-lives greater than one day)

Name	Symbol	Atomic number	Range of mass numbers: Lowest	Highest
Hydrogen	H	1	1	3
Carbon	C	6	12	14
Potassium	K	19	39	41
Iodine	I	53	124	131
Bismuth	Bi	83	205	210
Radon	Rn	86	222	
Radium	Ra	88	223	228
Uranium	U	92	230	238
Plutonium	Pu	94	236	246
Mendelevium	Md	101	258	

As an example, bismuth has an atomic number of 83, and all atoms having 83 protons in the nucleus behave as bismuth, whatever the number of associated neutrons. The number of such neutrons may in fact vary from less than 110 to over 130, but the nucleus is stable only when 126 neutrons are present.[1] In all the other atomic types, or 'isotopes', of bismuth, the atomic nucleus is in different degrees unstable and the atoms undergo radioactive decay, emitting radiation when they do so. The rate of this decay varies very widely for the different radioactive isotopes of bismuth. Thus when the nucleus contains 125 neutrons the atoms are relatively stable, only a small percentage of them disintegrating in every 10 000 years. However, when the number of neutrons is less than 122 or greater than 127, half or more of the atoms undergo radioactive decay within a day (Table 1.2).

Table 1.2
Bismuth isotopes (atomic number 83)

Mass number	Half-lives	Emissions
189	Less than 1.5 s	Alpha
190–2	Less than 1 min	Alpha and positron
193–200	1 min – 1 hr	Alpha and/or gamma, and positron
201–4	1–12 hr	Alpha and/or gamma, and positron
205	15 d	Positron and gamma
206	6 d	Positron and gamma
207	38 yr	Positron and gamma
208	370 000 yr	Positron and gamma
209	– stable	–
210	Two forms (3 million yr	Alpha and gamma
	(5 d	Alpha, beta, and gamma
211	2 min	Alpha, beta, and gamma
212	1 hr	Alpha, beta, and gamma
213–15	7–45 min	Alpha, and/or beta, and gamma

The rate at which the atoms of any of these isotopes disintegrate is measured by the 'half-life' of their radioactive decay. The half-life indicates the length of time required for half the atoms in any sample of the isotope to decay, half the remainder decaying in the next half-life, and so on. In ten half-lives the number of remaining unstable atoms is thus reduced to about 0.1 per cent of the initial number; and the rate of emission of radiation due to their disintegrations is correspondingly reduced by a factor of one thousand from its initial rate.

Just as the 'atomic number' specifies the nuclear charge of an atom and so determines the chemical element to which the atom belongs, so the atomic 'mass number' specifies the number of particles, that is, the total of protons plus neutrons in the nucleus. This therefore defines the particular isotope of the element to which the atom belongs. Thus the stable form of bismuth, with its 83 protons and 126 neutrons, is referred to as bismuth-209. Each 'nuclide' or form of any chemical element is specified in this way, the name of the element

[1] Superscript numerals refer to bibliographical references on pp. 181–6.

indicating the charge and so the proton content of the nucleus, and the following number defining its mass and so its proton plus neutron content. (Colloquially, one may refer to 'bismuth-209'. In scientific writing, the mass number and if necessary the atomic number appear on the left of the atomic symbol, as $^{209}_{83}Bi$.)

For almost all elements of atomic number lower than that of bismuth, one or more nuclear forms are stable, the other isotopes of the element being radioactive. For these elements the number of such radioisotopes that have been detected varies from one in the case of hydrogen (hydrogen-3, which has acquired the nickname of tritium) to as many as thirty for caesium, although many of these have very short half-lives of less than a second. For all elements of atomic number greater than that of bismuth, however, all isotopes are radioactive, with half-lives ranging widely from fractions of a second to many millions of years.

Origin of different radionuclides

Primordial radionuclides

Some radionuclides decay so slowly that significant fractions still remain of the amounts that were present when these materials were originally formed. This formation, with that of many stable nuclides (of which the decay is even slower, being zero), is thought to have occurred some ten billion (ten thousand million) years ago, during the conditions of very high temperatures and neutron bombardment of exploding 'supernova' stars. Such material became scattered through the interstellar dust, and incorporated in our planet when it was formed five billion years later.[2]

A second group of naturally occurring radionuclides consists of the products of the continuing decay of three of these primordial ones: thorium-232, uranium-235, and uranium-238. When any radioactive atom decays, its nucleus ordinarily changes either its charge or its mass, so giving rise to a nuclide of different atomic number or mass. This daughter nuclide (an area of terminology which the principles of sex equality have not yet invaded) may itself be radioactive, and a decay chain then continues until a stable nuclear form is reached. The uranium-238 chain descends in this way through 14 steps to reach finally a stable form of lead.[1] Radium is one of this series of naturally occurring radionuclides which, in its isotope of atomic weight 226, has the sufficiently long half-life of 1600 years for appreciable quantities of it to be present in uranium-bearing ores.

Cosmogenic radionuclides

A third group consists of those radionuclides formed by the impact of the neutrons present in cosmic radiation on atmospheric or terrestrial materials. A small fraction of all neutrons which 'hit' atomic nuclei become captured by these nuclei. The resultant increase in nuclear mass may create a radioactive form of the element irradiated; and, for example, significant amounts of the radioactive carbon-14, with a long half-life of over 5000 years, are constantly being formed by the capture of neutrons in the stable nitrogen 14 of the atmosphere.

It is worth noting that this group of naturally occurring radionuclides will include even plutonium-239, which is commonly thought of as being only of man-made origin. It is however formed also in small quantities as a result of the penetration of cosmic ray neutrons into uranium-bearing strata.

Artificially produced radionuclides

To these naturally occurring radioactive materials must be added the much greater variety of radionuclides which, as described later, can now be formed either by neutron or other particle bombardment of stable materials in cyclotrons or other devices, or which are formed by the fission of nuclear fuels in reactors or in the explosion of atomic weapons.

Radioactive emissions

The types of radiation emitted by atoms in the course of their radioactive decay were studied in detail by the Curies, by Rutherford, and by many others, and three principal types were identified. In one form, alpha radiation, heavy charged particles (each consisting of two protons and two neutrons) are ejected from the nucleus of the decaying atom. In the second form, beta radiation, the discharge is of particles of only very small mass but with a positive or negative charge – positrons or electrons respectively. The third type, gamma radiation, which is commonly associated with alpha or beta emission, involves a discharge of energy rather than of mass from the nucleus. Gamma radiation is an electromagnetic radiation similar to x rays, and occurs with a range of wavelengths and of energies which overlap those of the radiations produced by the x-ray tube (Fig. 1.1).

Table 1.3

Composition and approximate penetration of common types of ionizing radiation

Type of radiation	Constituent particles	Charge (per particle)	Mass (per particle)	Penetration (water or tissues) – varies wth energy
Alpha	Helium nucleus (2 protons + 2 neutrons)	+2	4	hundredths of a millimetre
Beta	Electrons	−1	(1/1840)	millimetres
	or Positrons	+1	(1/1840)	millimetres
Gamma	Photons	0	0	centimetres
Neutron	Neutrons	0	1	centimetres
X ray	Photons	0	0	centimetres

These three forms of radiation, although they all cause ionization, have very different characteristics which are determined by their mass, charge, or energy (Table 1.3), and which affect their interaction with biological material.

The emission of alpha radiation occurs mainly during the decay of the heavier atoms. Except in the case of a small number of primordial radionuclides of very long half-life, it is infrequent in nuclides of atomic number lower than that of bismuth. Alpha particles are slowed rapidly in their passage through water or body tissues. The total length of their path, from the radionuclide from which they are emitted to the point where they cease to cause ionization, varies with their initial energy but rarely exceeds one-tenth of a millimetre. Their initial energy is therefore fully absorbed in causing many ionizations over a very short distance. The biological importance of radiations having this rapid loss of energy with distance, i.e. with a high 'linear energy transfer' (LET), is discussed later (p. 96).

The passage of beta radiation through tissues causes tracks of ionization which vary in length according to the initial energy of the particles, but rarely exceed a few centimetres. In consequence, radioactive materials which emit beta radiation only, and which are outside the body, do not irradiate any of the deeper lying body organs. Radiation from such materials within the body is largely restricted to the organs in which these materials are concentrated or to their immediate vicinity.

Gamma radiation, however, having, like x rays, no charge or significant mass, is transmitted much more readily through body tissues. For the weakest gamma radiations the intensity is halved by every few centimetres of tissue, but, for the most energetic, by some tens of centimetres. Gamma radiation accompanies alpha or beta radiation in the decay of many radionuclides, but may also be released when the decay is associated with internal rearrangements of the nuclear structure without emission of either of these particulate radiations.

Some radionuclides decay with less usual forms of ionizing emissions, for example with discharge of x rays, protons, or neutrons. A few isotopes of the elements of highest atomic number may also disintegrate by undergoing spontaneous nuclear fission, the fission being accompanied by emission of energetic neutrons and gamma radiation.

EARLY MEDICAL USE OF RADIOACTIVE MATERIALS

In medicine, radioactive materials were at first of value only in treatment. For this purpose, radium was a convenient source since the gamma radiation emitted during its decay penetrated to deep body tissues more efficiently than the most energetic x rays that could be produced with the early forms of equipment. Radium was available, even if only in small quantities, in the ore pitchblende, the radium-226 being formed continuously by decay of the uranium-238 in the ore. The long half-life of this radium isotope, of 1600 years, ensured an effectively constant rate of emissions from a radium source, in contrast with the highly temperamental behaviour of early x-ray tubes. The slow rate of its decay involved the use of an appreciable mass, tenths of a gram or more, of this rare material to maintain the number of disintegrations per second required to produce a high radiation dose rate. However, when such quantities were available the use of a well-shielded capsule containing a chemical compound was much simpler than the generation of x rays in hospitals which at that time commonly had no regular electrical supply, and which depended upon hand or pedal operated generators of the venerable Wimshurst type. (In the running of these, one 1904 textbook of radiology[3] advocated the use of a 3 amp current and a 10 inch spark gap, which must have contributed an element of drama to the radiology of the day.)

Moreover, radium had the advantage that it could be introduced within the body, in tubes or needles, in such positions that it would cause intense local irradiation throughout a tumour without the risk of damaging the skin, which often limited the therapeutic dose that could be given with x rays.

When it was used in this way, however, the problem was to ensure that the tumour, and the areas into which it might be spreading, were adequately and uniformly irradiated. It was a disadvantage also that, if the needles had had to be implanted deeply within body tissues, a second operation was usually needed to remove them after an appropriate radiation dose had been delivered.

This difficulty, at least, could be avoided by using one of the daughter products of radium decay, the radioactive gas radon. Very small quantities of this gas were sealed into small metal 'seeds', and a number of these radon seeds were implanted throughout the tissue needing to be irradiated. In view of the short half-life, of less than 4 days, of the radon-222 isotope that was used, the seeds could be left in place, the required radiation dose to the surrounding tissues being delivered during the complete radioactive decay of the radon inserted. Moreover, by placing a number of these seeds in a suitable pattern, it was easier to obtain a uniform distribution of radiation exposure than when radium needles were used.

Both with radium and with radon, the thin metal containers in which they were used were necessary, not only to prevent the escape of any radioactive material, but also to screen off or absorb the alpha radiation which both radionuclides emit, which would otherwise cause very intense local irradiation at the surface of the container.

The difficulties of obtaining effective and reasonably uniform radiation of cancers or other abnormalities, without undue damage to skin or adjacent tissues, were to be substantially reduced when new radioactive materials of more appropriate properties could be artificially produced, and new techniques of irradiation developed, during the second half of the present century. The problems were already well understood and increasingly tackled, however, during its first fifty years, although not without many early and tragic failures.

THE RECOGNITION OF RADIATION INJURY

The scientists and their assistants were the first to suffer from over-exposure to x rays, and to recognize the damage that ionizing radiation could cause. The early investigators were usually unaware of the importance of avoiding even small but repeated exposures to radiation or to radioactive materials, and would sometimes hold their hand before the x-ray tube to demonstrate the image of its bones on a fluorescent screen, or merely to test whether the tube was emitting satisfactorily. And Becquerel himself in 1901 carried a tube of radioactive material in a waistcoat pocket until he noticed reddening of the underlying area of skin. (Then, characteristically, he repeated the practice, using tubes with and without radioactive material in them, to be sure that the effect was due to radiation.)

The earliest observations of reddening and inflammation of the skin in 1896 were soon followed by instances in which cancer was seen to develop in frequently or heavily irradiated skin areas; and the tumour has been mentioned already which was reported as early as 1902 on the hand of a maker and demonstrator of x-ray tubes in Hamburg.

With the rapid adoption of x rays in diagnosis, and particularly also at the high doses used in the attempted cure of a variety of medical conditions, such tragedies became more frequent, both in medical staff and in patients treated by x rays or radium. The monument in Hamburg to the early 'martyrs of radiation' records 169 deaths amongst radiologists alone. The long interval that commonly elapses between radiation exposure and the development and detection of any resultant cancer, however, delayed the full recognition of the dangers of the high doses that were involved in the early forms of radiotherapy and in the repeated exposure of radiologists. It became clear, however, that such exposures could not only cause severe structural damage to the tissues, evident microscopically in inflammatory changes and subsequent fibrotic scarring. They might also occasionally give rise to cancers developing within these grossly damaged and fibrosed areas of tissue. Moreover, the use of x rays of increasing energy and tissue penetration gave evidence not only that cancer might be induced in skin, but that it might occur in almost any body organ or tissue that was heavily irradiated and damaged in this way. The attempted treatment at this period of many non-malignant conditions showed that such cancers were new developments that had resulted from the radiation, rather than from a recurrence of a previously treated cancer.

The frequency with which radiation damage occurred – and it is reported that 170 cases of injury were described within 5 years at this period – will have resulted from a number of circumstances. The radiological use of x rays and of radium spread world-wide within two years to centres in which there was no previous experience of any comparable technique. The investigators and early radiologists had had no warning of such severe effects from their experience with the non-ionizing forms of radiation that had been studied or used, although the skin burns from over-exposure to ultraviolet and infra-red radiations were familiar. The greater initial use of fluorescent screens than of 'skiagrams' registered on photographic plates will have exposed both patients and radiologists to higher doses; and, although the need for full dark adaptation of the eye in examining the screen was early emphasized by Antoine Béclère and others, fluoroscopy was still sometimes conducted in sunlit rooms, using an increased brightness of the image and so an increased exposure of the patient. Moreover, in treatment little was known initially about the amounts of radiation needed for effective therapy, or the importance of minimizing the exposure of areas which were not being treated, although an early American radiologist treating a breast cancer in 1896 does record having screened the rest of the body from radiation using lead from Chinese tea chests.

THE DEVELOPMENT OF RADIATION PROTECTION

But even though the possibilities of serious harm were soon recognized, a major problem lay in estimating numerically the amounts of radiation exposure that were occurring. Early types of x-ray equipment were very variable in their radiation emission from day to day, and from one machine to another. Until reliable estimates could be made of the doses that were being delivered to skin and, even more difficult, to deeper tissues, no adequate protection measures could be developed, since information could not be obtained, or shared between different centres, of the tissue doses that had caused serious injury.

This difficulty was clearly understood, and many attempts were made to 'calibrate' or standardize x-ray equipment in terms of the quantity of radiation emitted under different conditions of operation. 'Quantitometers' were devised which depended on measurements of relative transmission through different metals, discoloration of crystals, fluorescence produced in suitable materials, or darkening of photographic plates. The very variety of the B-, D-, E-, F-, H-, and X-units used by various authors indicates the inadequacy of each. Even the human skin, a measuring device which was only too readily available, proved too variable in its reddening on exposure to radiation for the 'erythema dose' to have had more than a temporary and approximate value.

Technical difficulties also delayed the use of a direct physical measure of the radiation effect which was likely to be of biological importance, namely the amount of ionization produced. In principle, the ionization in a volume of gas could be estimated by collecting the total electrical charge carried by the charged ions, on electrodes mounted in the chamber containing the gas. Indeed, Villard in 1908 suggested basing a unit of x radiation on the quantity of this radiation which liberated one (electrostatic) unit of electricity by its ionizing effect on one cubic centrimetre of air, under specified conditions of temperature and pressure. It was to be 20 years, however, before this suggestion could be adopted internationally as a practical measure of radiation, either from x rays or, 10 years later, from radioactive substances. It was later still before the measure of ionization in air could be properly interpreted in terms of the corresponding ionization produced, or the energy delivered by this ionization, in body tissues by radiations of different energies. The units of radiation dose that are used today, however, derive from developments of Villard's proposal, since they are based on the amount of energy delivered by ionizations to a unit mass of tissue or other medium, and on the biological effectiveness of this amount of energy for radiations of different type and LET.

Meanwhile the urgent need to control the uses of ionizing radiation so that patients and medical staff should be properly protected led to detailed study of the techniques of radiotherapy, and, increasingly, to laboratory investigations into the ways in which radiation caused harm. National advisory committees on protection in radiology first published recommendations for good practice in Germany in 1913 and in Britain in 1915, and the importance of such precautions was being emphasized in the USA in 1916. The 1914-18 war, while impeding progress and international communications, led to a great increase in the use of x rays, particularly in detecting fractures of bone and in locating shrapnel or bullets in the tissues, and further demonstrated the need for care during the widespread use of rather primitive mobile x-ray equipment under field conditions. By the early 1920s, however, the radiological professions of a number of countries had adopted quite detailed recommendations for the proper management of x-ray and radioactive sources, of x-ray rooms, and of staff and patients during examination or treatment.[4]

These recommendations provided useful general guidance on ways to avoid unnecessary radiation exposure, but they could not give any consistent measure of the amount of exposure that could safely be tolerated, or indicate a reliable means of estimating it.

In 1925, however, an International Congress of Radiology was established and, at its first meeting in London, discussed the urgencies of radiation protection. Two of its decisions were to be of increasing importance in subsequent years.

Firstly, it arranged for a review of the numerous techniques of estimating radiation exposure, and the choice of a sound and reliable way of making this estimate. It was recognized that biological damage depended ultimately on the impact of energy delivered to cellular structures as a result of the ionization produced in the tissues. The amount of ionization that was caused in a volume of air by a given radiation exposure could now be measured reliably, on the basis already described. The unit adopted for estimating x-ray exposure, appropriately named the röntgen, was therefore defined in terms of the production, in a stated mass of air, of ions carrying a unit quantity of electricity, of positive or negative charge. The international acceptance of this unit, at the second congress in 1928, provided the necessary link between the measurable quantity of radiation to which the body was exposed, and the observed occurrence of minor or major degrees of tissue damage. This was an essential step in converting the previous, necessarily rather generalized and qualitative, recommendations on radiological protection into the more quantitative guidance on the amounts of exposure which should be avoided to prevent the occurrence of harmful effects, or to prevent their occurrence with undue frequency. The group of scientists set up to review the problems of radiation measurements was established by the congress as a standing International Commission on Radiation Units and Measurements (ICRU) to keep these problems under review in the light of developing knowledge.

The second important field which the congress arranged to review was that of the various national rules or recommendations on radiation protection. It was appreciated that it would clearly be valuable if recommendations of general application in radiological work could be agreed internationally, and could be kept under surveillance by a body of international scientific standing. A survey was therefore made of the protection procedures adopted in various European countries and in the USA, and particularly of the recommendations that had been developed by this time in Britain and in Sweden. As a result, the 1928 congress in Stockholm issued a set of quite detailed recommendations, and established a second commission, known later as the International Commission on Radiological Protection (ICRP), to keep such radiation protection recommendations under review.[4]

Within the first fifty years after the discovery of x rays and of radioactivity, therefore, a substantial body of knowledge had been built up and reviewed internationally on the properties of ionizing radiations and on valid methods of measuring amounts of exposure to them. The value of radiation in treatment was established and its usefulness in many diagnostic procedures was obvious. It was clear that large or repeated exposures causing gross tissue damage might sometimes cause cancer in the tissue irradiated, but it was not known from clinical evidence or from laboratory studies whether lower doses might have these effects, or whether tissue repair processes would reverse the effects of such low doses, as was known to occur with low doses of other harmful agents.

The next ten years were to see a rapid increase of knowledge of the effects of radiation from medical, genetic, and other radiobiological studies, but an even more rapid extension of the number of ways in which people were exposed to ionizing radiation.

2

NEW SOURCES OF EXPOSURE: ARTIFICAL RADIONUCLIDES

For all time, mankind had been exposed to radiation from natural sources. For fifty years, the use of x rays and naturally occurring radioactive materials in medicine had added to this exposure. Now, however, in the 1940s, the number of possible sources of human exposure was greatly increased. This followed essentially from two discoveries of effects caused by neutron irradiation. Neutrons could be used to create radioactive forms of the common chemical elements; and they could cause nuclear fission in certain of the heavier elements.

CREATION OF ARTIFICIAL RADIONUCLIDES

In 1934, Enrico Fermi found that various chemical elements, when irradiated with a stream of neutrons, became radioactive. Thus when stable iodine was irradiated, a radioactive substance was produced which decayed with a half-life of about half an hour, and could not be separated from stable iodine in any of its chemical reactions. It became clear that a new and previously unknown form of iodine had been created. One neutron had been incorporated into the nuclei of the stable iodine atoms, raising their mass number from the normal value of 127 to 128, rendering the nuclei unstable (Fig. 2.1). Since the addition of a neutron involved no change in nuclear charge, however, the chemical behaviour of the new radionuclide was still that of iodine. (The iodine-128 then decays by loss of an electron to a stable form of the gas xenon.)

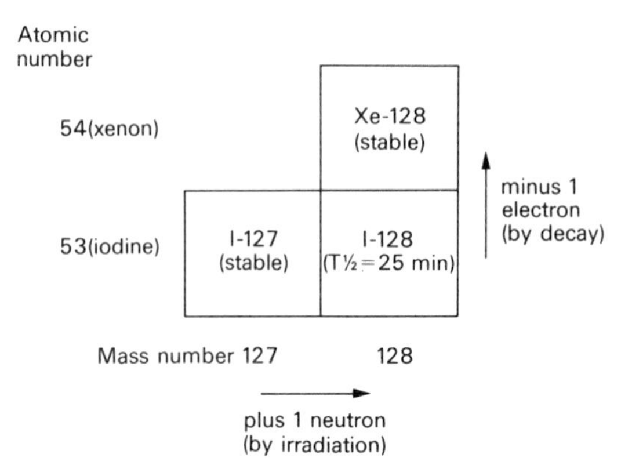

Fig. 2.1. Transformation of iodine-127 following neutron irradiation (to iodine-128, which decays to xenon-128).

When phosphorus was irradiated with neutrons it also became radioactive. Now, however, the radionuclide created did not behave like phosphorus, but it did have the same chemical reactions as silicon, an element of atomic number one less than that of phosphorus (Fig. 2.2). In this case the addition of a neutron to the phosphorus nucleus had displaced a proton from it, producing a new radionuclide of mass equal to that of the irradiated phosphorus but of lower charge. (This silicon-31 decays back to the original phosphorus-31 by loss of an electron.)

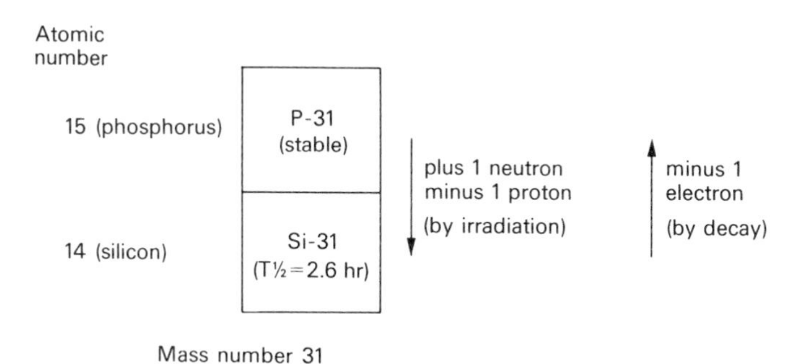

Fig. 2.2. Transformation of phosphorus-31 following neutron irradiation (to silicon-31, which decays back to phosphorus-31).

The synthesis of artificial radionuclides had in fact been detected by Irène Curie and Frédéric Joliot in the previous year as a result of the alpha particle irradiation of various substances. The use of neutrons, however, increased greatly the variety of radionuclides that could be formed, and considerably increased the yield of these materials. This yield was still minute in mass, and would not have given much satisfaction to the alchemists, who had been trying to transmute the chemical elements for long enough. The amounts were, however, sufficient to allow the chemical and metabolic behaviour of many important chemical elements to be examined with great sensitivity by measurement of their radioactive forms. This early use of such chemical 'tracers' foreshadowed the whole development of nuclear medicine, with its present range of 'radioisotope' tests and treatments, and with the highly sensitive and important methods of measuring very low blood concentrations of different hormones and metabolites in the detection or study of disease.

TRACER STUDIES

The increasing availability of radioactive forms of essentially all the common chemical elements resulted in an enormous and rapid increase in our understanding of the processes that take place in the body in health and disease. Much was, of course, known already about the chemistry of these processes, and about the importance of iron, calcium, carbon, iodine, and other elements in the normal metabolism of body tissues. Many such elements, however, and the compounds in which they are incorporated, are present at such low concentrations in the blood or in the tissues that their detailed study by normal chemical methods is difficult or impossible. Moreover, the concentration of some of the most important of these compounds is so accurately controlled by body processes that the nature of this control and its failures in disease cannot be examined by chemical methods. Any administered amount of the substance which caused a measurable change in its concentration in the blood would grossly alter the controlling processes which were being investigated.

Measurement of the concentration of a substance by means of its radioactivity, however, can allow a sensitivity a million or more times greater than that attainable by chemical means. Studies of metabolic behaviour can therefore be carried out by methods which involve no risk of distorting the very processes which they are intended to study. The normal human thyroid, for example, takes up about 30 per cent of the daily intake of iodine, with which to form the iodine-containing thyroid hormones. Its efficiency of uptake can readily be measured if, to this daily intake of 100 micrograms of stable iodine, is added one millionth of a microgram of iodine in the radioactive form of iodine-132, an amount which will not itself alter the gland's efficiency of uptake.

The use of artificial radionuclides as tracers thus provided wide scope and great sensitivity in biochemical studies, with radiation exposures which were in most cases so low that they would be most unlikely either to cause harm, or to disturb the processes investigated.

Such investigations had in fact been made before any radionuclides had been artificially produced. George de Hevesy had been given by Rutherford in 1912 the task of separating one of the radioactive decay products of radium from the lead in which it was present in mining samples. When all available chemical methods had failed to effect any separation, Hevesy concluded that the decay product must in fact be a form of lead itself. He then proceeded to study various chemical and physical aspects of the behaviour of lead, by using this radioactive isotope of the metal. He first examined the engineering problem of the transfer of metal between two bearing surfaces, by amalgamating his radioactive lead into the metal of one surface, and detecting, with high sensitivity, its gradual transfer into the metal of the other, as the bearings continued to engage with each other. The detection of tracers in the lubricant of engines today provides a continuous record on the wear of engine parts, with high accuracy and at low cost.

Hevesy then studied the sites of uptake of lead, and later of bismuth, into plants from the soil, by adding naturally occurring radioisotopes of these elements to the soil and measuring the radioactivity incorporated into different parts of the plant system. With his co-workers in Copenhagen, he also compared the body organs in which these elements were retained, and the rates with which they were excreted, after their injection in small amounts into animals.

No naturally occurring radioactive isotopes were available, however, of elements which played any important part in normal metabolism. While studies of the biological behaviour of lead, bismuth, and various other elements in the early 1920s showed the power of tracer methods, the value of these methods was not fully realized until the artificial radionuclides started to be produced in increasing variety and quantities.[5]

PRODUCTION OF RADIONUCLIDES

Fermi had used a source of neutrons generated by the impact on beryllium of alpha particles emitted from radon gas. Within a few years, however, electrical equipment was devised which could supply beams of neutrons or other particles at much greater intensity, and these cyclotrons could produce considerably higher yields and a greater variety of radionuclides. It was not until the mid-1940s, however, with the development of nuclear reactors, that substantial quantities of artificial radionuclides became more readily available, which could supply the needs of medicine, industry, and many fields of research.

Radionuclides can be produced in reactors in two ways. Firstly, there is a very high flux of neutrons in the interior of the reactor. Radioactivity can therefore be induced in materials that are placed within the reactor core in the same way as in Fermi's earlier work, but in much larger amounts. And secondly, as discussed in the next chapter, large amounts of radioactive products result from the fission of uranium in the reactor. A wide range of biologically important radionuclides can be separated in quantity from this mixture of fission products.

MEDICAL APPLICATIONS OF ARTIFICIAL RADIONUCLIDES[6-8]

The medical applications of radionuclides depend essentially upon two of their properties: firstly, that these nuclides are chemically identical with the corresponding stable forms, so that radioactive calcium or phosphorus will be metabolized in the body in the same way, and at the same rate, as normal stable calcium or phosphorus; and secondly, that their presence and amount can be measured with such ease and sensitivity by simple counting of the frequency of their radioactive emissions, whether determined in samples of body fluids or excretions, or by gamma radiation transmitted through the body wall.

These features allow them to be used in several important ways:

1. In estimating the speed or efficiency of the function of a body organ, when this is reflected in the rate with which the radioactive form of an

element, or any compound containing it, is concentrated into the organ or discharged from it. Similar principles apply to determinations of the volume of blood or other body fluids through which a tracer substance becomes distributed, and to the excretion rate of different materials in health and disease.

2. In examining the sequence of chemical transformations of an administered substance, and the speed and efficiency with which they are carried out, or the partitioning of such metabolites between different body tissues. A similar principle determines the value of radioactive labelling agents in 'radioimmunoassays', in which the chemical distribution of the labelled compound allows very low concentrations of biologically important substances to be measured, in blood or other samples withdrawn from the body, as described below (p. 22).

3. In examining the spatial distribution of labelled materials throughout the body or within body organs, particularly by determining the outline of organs or the presence of tumours in which they become concentrated, by 'scanning' over the body surface to map out the penetrating gamma radiations emitted from the underlying sites of radionuclide retention.

4. In measuring the concentration, or detecting the presence, of various chemical elements in human tissues or blood samples, by rendering these elements radioactive as a result of neutron irradiation.

5. By utilizing the selective concentration of certain radionuclides in cancer or other tissues, to deliver intense irradiation locally to these tissues, particularly by beta or other radiations of low penetration, with good prospects of destroying the abnormal tissue without undue irradiation of any other body tissue. Moreover, certain radionuclides can be used as substitutes for the naturally occurring radium or radon in radiotherapy, not only because of their greater availability, but also because radionuclides can be selected that have emissions of a type and energy appropriate to the thickness of tissue penetration needed for full tumour irradiation. This applies to the replacement of radium in its use both as an external source of radiation, and when inserted into the body in sealed containers.

It is worth illustrating briefly the scope of these various uses of radionuclides in medicine, and of applications in other fields. In this context, it is important to appreciate that, as discussed in Chapter 5, the amount of radiation to which the body is exposed in a range of radionuclide tests and investigations is similar to that from a range of typical diagnostic examinations by x ray.

Measurement of organ function

The measurement of the function of an organ is well illustrated by the ways in which the efficiency of the lungs, or of individual segments of the lungs, can be examined. Firstly, the capacity of the airways to deliver air normally to any part of the lung can be measured, and any local blockage detected, if a small amount of a radioisotope of a gas such as krypton or xenon is mixed with the air to be breathed, and the radioactivity over different areas of the chest measured during and at the end of an inspiration. These gases are normally present in air in their stable forms, and serve to assess the normality of regional lung ventilation in this way.

Secondly, the efficiency of the blood circulation through any part of the lung can be measured by testing also the speed with which this local intake of radioactive gas is removed by the blood during a short period while the breath is held. To ensure that this measure of removal rate is valid for oxygen also, determinations have been made with radioactive oxygen itself. The only available radioisotope of this element, however, has the very short half-life of about half a minute, so that the tests have to be made in close proximity to a cyclotron in which the oxygen-19 is being continuously produced, and piped to the breathing equipment.

Similar tests of the lung circulation can also be made by intravenous injection of a solution of a radioactive gas, and measurement of the speed with which radioactivity appears in different parts of the lung.

It is evident that comparable measurements of clinical importance can be made on other organs: on the blood supply to the kidney, by the rate of arrival of radioactively labelled substances which become concentrated in it, and on its excretory function by the rate of their subsequent removal; on the activity or over-activity of the thyroid, by the rate and degree to which it takes up iodine or other substances which it normally concentrates; or on the performance of the spleen in removing damaged red blood cells from the circulation, as examined by injecting labelled and heat-treated red cells. Similarly the presence and amount of any bleeding into the intestine can be assessed by removing, labelling, and reinjecting a sample of the patient's own red cells, and measuring the rate of loss of radioactivity from the body, over a period during which it is known that the labelling radionuclide does not become detached from the red cells.

It may also be important to know the amount of blood in the body, or the volume of tissue fluids through which biologically important substances such as sodium, potassium, or albumen are distributed. The size of such 'distribution spaces' is readily measured if a known amount of the relevant substance is injected into the bloodstream in labelled form and a known volume of blood is removed and counted, after allowing the time necessary for uniform mixing throughout the blood and tissue fluids. Any reduction in such distribution spaces may give valuable evidence of disease which is not obtainable by other means, since the amount of the substance per unit volume of blood may be accurately maintained by body processes at a constant level, even though its distribution space is greatly reduced.

Alternatively, in the diagnosis of anaemias caused by an increased rate of red cell destruction, a known volume of blood is removed, labelled, and reinjected.

The average survival-time of red cells in the circulation, which is normally of about four months but may be greatly reduced in these anaemias, can be determined by the proportion of red cells which are still found to be labelled in samples withdrawn at intervals after the original injection. It would seem obvious to label the red cells with a radioisotope of iron, since iron is a normal constituent of these cells. In fact, however, iron is efficiently reutilized in the formation of new red cells when old ones are broken down, so that the new cells would become equally labelled. Instead, therefore, a radioisotope of chromium can be used, since a radiochromium label can be firmly attached to the cell, but is not reutilized when the cell is destroyed.

Finally, just as estimates can readily be made of the efficiency with which different substances are excreted from the body in various diseases, the absorbtion of materials from the gut can be assessed, and has proved to be of value both in the investigation and in the diagnosis of certain diseases. Studies of the efficiency with which calcium, iron, and fats are taken into the body from the diet have been important in this way; and certain types of anaemia which are due to defective absorbtion of vitamin B_{12} have been studied in detail, and are diagnosed, by oral administration of this vitamin labelled with a radioisotope of cobalt, an element which is a normal constituent of the vitamin.

Study of chemical transformations

The chief use of radionuclides in examining the course of chemical transformations in the body has been in studying the abnormalities involved in metabolic diseases, rather than in routine clinical diagnosis. Important experimental work has been done, both in man and in animals, to determine the substances derived from the diet which contribute to the formation of more complex tissue components, or the steps by which new drugs are broken down and excreted from the body. If the substance or the drug is synthesized with, for example, a radiocarbon atom stably incorporated within its structure, the presence of this radionuclide in any compound found in subsequent urine, blood, or tissue fluid samples shows that compound to have been derived, at least in part, from the administered material.

Information is also obtainable in this way about the timescale on which such chemical changes take place. Thus, for example, the speed of hormone formation in the thyroid gland can be inferred from the rate at which radioiodine appears in samples of the thyroid hormones present in the blood, after radioiodine has been given by mouth in the simple chemical form of iodide.

A major value of the labelling of chemical substances, however, has been in the powerful and important techniques of radioimmunoassay.[9] These methods depend on preparing an antibody which will selectively precipitate a hormone or other substance of biological importance from a blood plasma or other sample. For many such substances, the concentration in plasma is much too low to allow any direct estimate, by weighing or chemical determination, of the actual amount precipitated. The amount present in the sample can, however, be accurately estimated by the following ingenious method.

A sample of the hormone is prepared in pure form, and stably labelled with a suitable radionuclide – commonly one of the radioisotopes of iodine. This labelled hormone is then mixed with sufficient antibody to bind all the hormone to the antibody molecules. Graded amounts of unlabelled hormone are now added, so that increasing proportions of the labelled hormone become displaced from their binding to the antibody, as the labelled and unlabelled hormone molecules compete for the available binding sites. The proportion of the labelled hormone which remains bound to the antibody, for any given total amount of hormone present in the system, is readily measured by the relative amounts of radioactivity precipitated with the antibody or remaining in the solution. A calibration curve is thus obtained, relating the fraction of labelled hormone which is bound, to the total amount of hormone added.

When therefore a plasma sample containing an unknown amount of hormone is added to the same initial mixture of labelled hormone and antibody, the amount of hormone in the plasma can be deduced directly from the fraction of labelled hormone which remains bound to the antibody, since the calibration curve shows what fraction corresponds to what added amount.

In this way, with additional procedures for checking the stability of the labelling and the consistency of the fractionation, measurements can now be made routinely of the plasma or other concentrations of a very large number of substances of great diagnostic or investigative importance.

Imaging[10,11]

One of the most versatile and valuable developments of nuclear medicine consists in the delineation of body organs, using their natural capacities to concentrate different chemical substances, which can then be administered in radioactive form. In many circumstances these methods of radionuclide imaging replace or supplement the use of x-ray examinations, since they can reveal areas of functional failure within an organ which are not discriminated by normal x rays, and which may not be discriminated even by the refined methods of 'computerized axial tomographic' (CAT) scanning. Thus, a substance which is normally concentrated by the cells of an organ will not ordinarily be so concentrated in areas in which the cells are damaged by disease of the organ or replaced by deposits of cancer tissue.

The method was first used, in the early 1940s, to map the outline of the thyroid gland, and the functional state of swellings in the gland. Use was made of the high selective concentration of radioactive iodine in normal thyroid tissue, a radiation detector making measurements over the surface of the neck of the underlying radioactivity, and so giving a two-dimensional map of the sites and amount of thyroid activity.

This procedure has been refined and greatly extended in two ways. In the first place, the 'gamma camera' allows the amount of gamma radiation from all areas in its field of view to be recorded simultaneously, in the same way as other cameras register the light from different parts of the field. Images are thereby obtained much more quickly than by a counter scanning more slowly across the field, the camera building up an image, often with as much detail as appears on a television picture on the screen; and a series of such images can be examined while the concentration develops.

Secondly, the number of body organs of which the position, extent, and functional capacity can be determined has increased so much that there are now few organs or specific tissues that cannot be outlined in this way. This has been achieved by identifying different radionuclides, or compounds in which they can be incorporated, which become more or less selectively concentrated and retained in the different organs. Substantial advances have also been made in finding substances which are more highly concentrated in cancers of various types than in surrounding normal tissues, and which can therefore be used in detecting the presence or the spread of these forms of cancer.

The labelled chemical compounds, or 'radiopharmaceuticals', appropriate for the imaging of the different organs or tissues are chosen according to the substances normally concentrated in the organs, or from others which are found to behave in a similar way. The choice of the radioactive 'label' then depends on the stability with which it can be incorporated, either as a normal constitutent of the compound, as with cobalt in vitamin B_{12}, or as a chemically reactive substance like iodine or technetium which is unlikely to be lost significantly from the labelled molecule before the measurements are completed, and which does not alter the biological behaviour of the labelled substance. It has usually been easy to select radionuclides as labels which give an adequate number of counts to provide the necessary clinical information, without causing any undue radiation exposure to the organ or to the body as a whole. This is achieved by choosing a radionuclide with an appropriate type and energy of the radiations emitted, and with a half-life allowing enough activity at the time when concentration in the organ is complete and measurements are made, but no long persistence in the body once they have been made.

Finally, a type of mini-imaging has been used, from the earliest days of Hevesy's work, whereby the distribution of suitable radionuclides in specimens of tissue can be examined on a microscopic scale. A thin section of the tissue is covered with a sheet of photographic film, and time is allowed for any local areas of radionuclide concentration to produce a corresponding local exposure of the film. Such 'autoradiographs', in which the exposed film is seen superposed upon the stained microscopic section of tissue, can be informative in a few situations by indicating the metabolic behaviour, and therefore the identity, of local areas of abnormal tissue within an organ (Fig. 2.3).

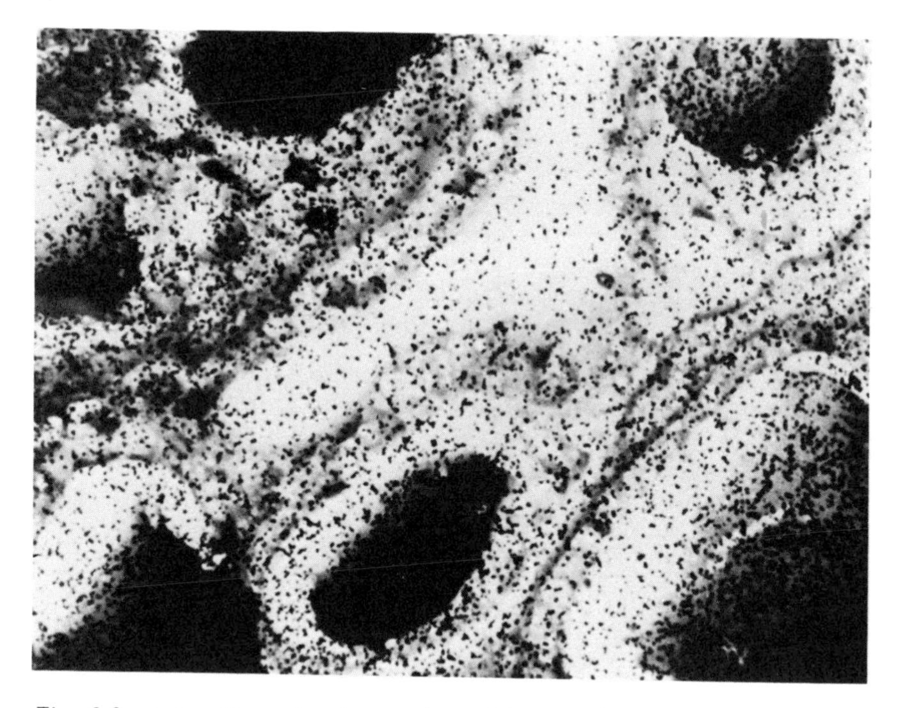

Fig. 2.3. Autoradiograph of a section of human thyroid cancer, indicating concentration of radioactive iodine in colloid-retaining follicles of the cancer tissue by the density of exposure (black grains) of photographic film overlying the positions of these follicles.

Neutron activation analysis

When neutrons of low energy are 'captured' in the nuclei of stable elements, the newly formed radioisotopes of these elements can ordinarily be identified by the types and the energies of the radiations that they emit, and by the half-lives of their radioactive decay. It is therefore possible to detect and measure very small amounts of various elements, by irradiating with neutrons the materials in which they are present, and recording the characteristics and the amounts of the radiations that are emitted subsequently.

This technique of neutron activation analysis has two useful applications in investigative medicine. Firstly, since its sensitivity is far greater for some elements than has previously been possible by chemical procedures, low concentrations can be measured in blood or tissue fluids of potentially toxic agents such as cadmium or arsenic, and some normal body constituents can usefully be measured in samples in this way.

Also, however, since the sensitivity of the method is so high, measurements can be made on the body or parts of the body, using neutron exposures giving tissue doses no higher than the equivalent doses received from many x-ray or radionuclide tests. This can be of considerable importance since, in a few diseases, danger results from changes in the amount of element such as potassium, nitrogen, or calcium in the body as a whole, and these changes sometimes cannot be assessed or even detected by examination only of blood or other samples taken from the body, or by any other methods.

In a few instances, therefore, estimates of the presence, severity, or cause of a disease can be made by the specialized techniques of brief irradiation of the body, or of a limb or other part of the body, to a dose of about 10 millisieverts (see p. 46) of low energy neutrons, followed immediately by a sequence of measurements in a whole body counter (p. 54) during the course of the transient radioactivity induced.

Radionuclides in treatment

In all forms of radiotherapy, and particularly in the treatment of cancer, the objective must be to deliver a high enough dose of radiation to all the abnormal cells to cause their destruction or arrest their growth, with a low enough dose to normal tissues to avoid unacceptable harm.

In the ordinary use of x rays, this objective is approached in several ways. First, the beam of x rays is planned to be wide enough to include not only the known position of the tumour but also a zone round it into which the growth may be extending. Secondly, the energy of the x rays used is sufficiently great for the 'depth dose' at the position of the tumour to be as high as possible relative to the dose to the skin through which the beam must pass. When x rays of low energy were used in the early days of radiotherapy, the progressive absorbtion of energy as radiation passed through the tissues meant that the skin dose was considerably higher than the tumour dose. Subsequent developments in therapy with x rays of higher energy, or sometimes with neutron or other particle beams, have greatly improved the ratio of depth dose to skin dose.

A high ratio of depth dose to skin dose can also be obtained in treatments with the high energy gamma radiation of a radioisotope of cobalt. A large amount of this cobalt-60 is needed for this purpose, partly to achieve the high dose rates required in therapy, and partly so that these high tissue doses can be obtained without the cobalt source needing to be dangerously close to the skin. The necessary amounts of cobalt-60 are obtained by neutron irradiation of the natural stable cobalt-59 in the core of a nuclear reactor. The resultant radionuclide has a conveniently long half-life of rather over five years, so that the dose rate from the source changes only slowly with time.

The problem of avoiding undue skin irradiation does not arise in the treatment of some cancers which occur in accessible positions where radioactive material can be directly implanted into the tumour tissue. Localized irradiation of this kind can now be achieved more conveniently and effectively than was possible with radium needles and radon seeds, by using the more flexible approach of inserting wire containing radioactive tantalum or 'seeds' of radioactive gold. Alternatively, in a few instances it may be valuable to inject directly into tissues through which cancer cells are spreading, a colloidal suspension of a suitable radionuclide. The use of colloid particles ensures that the material remains largely at the site of its injection. The use of a radionuclide emitting, for example, high energy beta particles but no gamma radiation ensures a relatively uniform radiation dose throughout the injected tissue without significant exposure at more than a few millimetres from this site.

A similar selective radiation of the tissues requiring treatment is occasionally used by injection of such high energy beta emitters in colloidal or insoluble form into body cavities of which the surface layers need to be irradiated, either because tumour tissue is spreading over these surfaces, or because the disability from certain types of arthritis may be relieved by irradiation of the linings of the joint cavities.

In all these instances, deep body tissues may be irradiated without exposure of the skin to radiation. The converse holds when tumours of the skin or of the eye can be effectively treated, without irradiation of deeper-lying tissues, by using applicators impregnated with radionuclides emitting beta particles of an energy appropriate to the thickness of the surface tissues which need to be irradiated.

In the procedures described so far, a radionuclide with appropriate half-life and energy of emissions is actually placed in positions which ensure the maximum dose to tissues requiring treatment, and the minimum to normal tissues. In a few instances, however, effective therapy is achieved by giving a radionuclide in such a chemical form that the body *sends* it to the tissues requiring treatment.

The most important example of this method is in the treatment of those forms of thyroid cancer in which the cancer cells retain the capacity of normal thyroid cells to concentrate iodine (Fig. 2.3). When this capacity is present in the cancer, or can be induced in sufficient degree by suppression of normal thyroid activity, the administration of large amounts of a radioactive isotope of iodine can result in the cancer tissue receiving very much higher radiation doses than other body tissues, and these doses commonly cause extensive or complete tumour cell destruction. Moreover, since the localization of the radioiodine and therefore of the high radiation dose is determined by the chemical behaviour of the cells of such cancers, the irradiation is delivered to all sites of tumour spread, whether they have been detected clinically or not. This makes it possible for thyroid cancers which cannot be completely removed surgically because of their local or remote spread, to be effectively treated in a way which is difficult for most other cancers, since the positions into which other cancers are beginning to spread often cannot be identified and treated. In addition, the sensitive detection of remaining sites of radioiodine concentration by external counting or chemical methods can be used to indicate, with considerable precision, whether the destruction of the functioning cancer tissue is complete or whether further therapy is needed.

A similar treatment with radioiodine is widely used for control of the much more common condition of simple over-activity of the thyroid gland. Here, however, the amounts of radioiodine that need to be given are much smaller, since the need now is to decrease moderately the activity of tissue which concentrates radioiodine highly, whereas with the cancer the need is to destroy completely all cells of a tissue which concentrates the radionuclide less strongly, and therefore receives a lower radiation dose per unit administered activity.

OTHER USES OF ARTIFICIAL RADIONUCLIDES

The non-medical applications of artificial radionuclides have been as varied and as valuable as those in medicine. They have been based on the same features of radioactive materials: the great sensitivity with which they can be detected, either when used as tracers or when induced in small amounts by neutron activation; their value as convenient sources of radiation under conditions in which the generation of x rays may be impracticable; or in some circumstances as a basis for the supply of energy.

Tracer applications

The tracer applications of radionuclides have, as in medicine, been used to study the way in which materials are transferred from one part of a system to another, and the rate at which this transfer takes place. The systems studied have ranged from the movement of tsetse flies around their breeding grounds, to the world-wide circulation of air masses in the upper atmosphere. Such studies have had a continuing importance in the development of agricultural and various industrial techniques, in methods of pest control, and in the understanding of large-scale air and water movements.[12, 13]

In some instances, the information is obtained by incorporating a radionuclide in the chemical structure of some of the material whose distribution is being examined. Thus, the best ways of introducing phosphate fertilizers into certain soils were tested by labelling the phosphates with phosphorus-32, and measuring the concentration of this radionuclide in growing crops, after putting the fertilizer into the soil at different depths or at different times, or using different sources of natural or prepared phosphates. In the same way a number of pesticides have been artificially labelled, to identify the extent to which any residues of the pesticide might enter animals or crops, or the likelihood that they would persist in the environment.

Radionuclide labelling has also been widely used to study the size of the territories, or the range of migration, of tsetse flies, black flies, and other vectors of disease that are rather too small to be caught and ringed in the conventional way. The uptake of a suitable gamma emitting radionuclide into the insect from a labelled material on which it feeds can be sufficient to define the range from the food source within which labelled insects are recorded. The same method has been used also to identify the main predators of the insects, with some assessment of the size of their appetites.

Similar principles are applied industrially, for example in the continuous registration of the wear of mechanical components, as already mentioned (p.18), or of that of blast furnace linings, by spiking the lining material with small amounts of cobalt-60 and recording the appearance of this radionuclide in successive batches of steel.

The meteorologists, however, would have had problems in labelling the upper atmosphere of the Northern Hemisphere sufficiently to allow its circulation to be investigated. They therefore took advantage of the period when this had been done for them, by the substantial discharges into these air masses of fission products from atmospheric weapon tests between 1954 and 1962. The observations obtained gave clear-cut evidence of the rate at which the lower, tropospheric, air circulated round the globe from west to east, as measured by the times of appearance of fresh fission products at different localities after any of the larger atmospheric tests in the same hemisphere.[14] Quantitative evidence was obtained also, for the first time, of the average period during which material remained in the upper, stratospheric, atmosphere before descending into the troposphere; the seasons of the year – mainly in spring – at which it did so; and the slow rate at which the stratospheric air from the two hemispheres mixed above the equator.

Large-scale studies on water movements took advantage also of the labelling of water masses during the same period by fallout, using particularly the tritium fallout on to surface waters during these years. This work, with parallel studies on the ratio of the stable isotopes of certain elements, has given information on the circulation of water at different depths in lakes and oceans; and also on the rate at which the 'ground-waters' below the surface of deserts or other land masses are replenished. This last point is important as determining whether the pumping of such water to the surface would yield a continuous supply, or would deplete the water layer. The presence or absence of an increased tritium concentration due to surface fallout would show whether the water was constantly circulating and being replenished, or was old and static and had not had any input from the surface during or since the period of heavy testing. The water tables closely below the Sahara do not appear, on these criteria, to have received any tritium from surface waters within a period of some twenty years. The deeper layers cannot have had any substantial replenishment since the last ice age, as judged by their dating by the naturally occurring carbon-14 and by a stable isotope of oxygen.

Neutron activation analysis

The analysis of materials using neutron irradiation, as described above (p. 25), has been used extensively in situations in which the normal methods of chemical analysis were not easily applicable: either because a rapid series of analyses was required; or because samples of the material could not be removed for analysis; or, most commonly, because chemical methods would be insufficiently sensitive.

Thus, for example, the amounts of aluminium in specimens of the ore bauxite can be rapidly determined, since neutrons induce radioactivity in the stable aluminium-27 by forming aluminium-28 which decays with a half-life of about 2 minutes. The grade of a series of ore samples can therefore be determined immediately from the counting rates observed and the mass of the samples.

Non-destructive testing by neutron activation is illustrated by a number of tests that have been used in detecting art forgeries.[15] Paintings, for example, using white paints of a zinc oxide basis are shown not to have been painted before the nineteenth century, since lead whites only were ordinarily available in earlier centuries; and a large variety of other chemical elements, identifiable by neutron activation without removal of paint from the picture, have been shown to characterize different periods or schools of painting by their presence in different pigments or as impurities in them. Moreover, the two-dimensional distribution of different paints over the surface of the picture, or in its invisible deeper layers of paint or priming, can be revealed in autoradiographs (p. 24) made following neutron irradiation – and may give further evidence of the likely date and source of paintings.

The high sensitivity obtainable by neutron activation analysis also has a variety of applications in biological, industrial, and forensic investigations, and particularly when small samples of material are available for analysis. Indeed, one of the earliest uses of the method was that made on the lock of Napoleon's hair, to settle the question of whether he had been poisoned or not. The chemical composition of the hair reflects to some extent that of the blood or tissue fluids at the time it is being formed, and arsenic was in fact found in parts of the hair first examined. The measurements were, however, said to be inadequate to determine whether the arsenic was attributable to the numerous contemporary medicaments that the Emperor was receiving, or to more sinister sources.

Radionuclides as sources of radiation

The first image of a welded structure was made in 1896 by x ray. A compact and portable source of penetrating radiation, which can produce radiographs of the structure of welds, metallic castings, alloys or pipelines, however, and reveal defects in them has a much wider field of application than any fixed or mobile x-ray equipment. Sources of cobalt-60 and other artificial radionuclides have been used in this way in the field, to detect cracks or inclusions in welds, unfilled cavities in metal castings, defects in pipelines, and unequal mixing of the constituent metals in steels or alloys.[16]

These applications require a highly penetrating gamma radiation to penetrate the metallic structures efficiently. At the other end of the scale, however, radionuclides with emissions of low penetration have been used to monitor continuously the thickness of paper or other material during its production, the amount of radiation passing through the paper giving the basis for a sensitive thickness gauge.

The gamma radiations from cobalt-60 and caesium-137 sources have been used in many countries in the preservation of food. The use of such methods is, of course, of particular importance in developing countries with hot climates and scanty facilities for refrigeration, and where fixed radionuclide sources are easier to operate and maintain safely than x-ray sterilizing equipment. Under these circumstances particularly, irradiation can considerably reduce post-harvest losses and wastage of fresh or dried foods by killing or sterilizing insects or other parasites, or by inhibiting the sprouting of root crops. The radiation doses required to produce these effects, and the safety of the foods so treated, have been studied by teams set up jointly by the Food and Agriculture Organization and the World Health Organization. In tropical developing countries, food preservation by radiation must be reducing considerably the risks and hardship not only from food infection, infestation, and other sources of wastage, but also from the need for high levels of chemical preservatives, fumigation, and other traditional methods of doubtful safety.

As well as the preservation of food, the sterilization of dressings and other medical supplies, and the established experimental use of radiation in causing mutations in seed to produce resistant crops, one of the more promising forms of tropical disease control depends also on the field use of radiation sources. Insecticides are effective in reducing considerably a local population of disease-carrying insects, but not in completing the elimination of, for example, the tsetse fly, which transmits sleeping sickness. If, however, male flies are sterilized and released in the area, it seems likely that this elimination can be completed and the area kept free of significant reinfestation by occasional subsequent such releases. It has been found that male flies can be bred in captivity, and when released after exposure to a high dose from a cobalt-60 or other appropriate source, will mate with normal females in the environment but the union will not be blessed with offspring, so that the free population becomes progressively reduced as the programme continues. The method is ineffective against heavy infestations, and so requires the initial attack with insecticides. Once the infestation is reduced, however, elimination has been shown to result from continued release of sterilized males, at least for one riverine species tested; it is, moreover, specific to the insect vector attacked, rather than for the range of species that may be vulnerable to the insecticide.[17]

Nor does the method require a keen-eyed entomologist catching and sterilizing males only. The flies can now be bred with a sex-linked resistance to certain lethal agents so that only males survive the breeding process and no females, even sterile ones, are released.

Radionuclide sources, but this time weak ones, are used also in luminizing materials for watches, clocks, and other instrument dials, and as small continuous 'beta light' sources, such as those which mark the position of light switches or are incorporated in exit markers or smoke detectors. The property of radiation sources in causing ionization of the surrounding air is utilized also in 'starters' for fluorescent lamps, voltage stabilizers, and various forms of trigger tubes in electrical appliances or excess voltage protection devices. In all of these domestic and small industrial appliances, the objective is to provide a very limited local field of radiation, with no escape of penetrating radiation. For this reason, sources of beta radiation from tritium (hydrogen-3) or promethium-147, or in some cases of alpha radiation from americium-241 incorporated in foil, are used. Thorium is also widely used in vapour discharge lamps and in various types of vacuum appliance.

The radiation exposures of people from these and other domestic sources of radiation are in general very small, but need to be assessed (p. 88).

Radionuclides as energy sources

A number of methods have been developed for converting the energy released in radioactive decay into electrical energy, and 'batteries' generating electricity in this way have been used in space vehicles, in lightships or buoys, and in cardiac pacemakers: in situations, in fact, in which a continuous output is required for long periods of time, from a source which should be reliable without need for maintenance, surveillance, or adjustment.

3

UTILIZATION OF NUCLEAR FISSION

DISCOVERY AND BEHAVIOUR

When Fermi discovered in 1934 that radionuclides could be artificially produced by neutron irradiation of various chemical elements, he had included among the substances which he irradiated uranium, which was then the element of highest known atomic number, 92. Radionuclides of several different half-lives were produced. These were at first assumed to be isotopes of unknown elements of higher atomic number than uranium, produced by the addition of neutrons to the uranium nucleus, since they did not correspond to any of the likely immediate decay products of uranium.

As the studies of various workers proceeded, however, the number of these neutron-induced radionuclides became embarrassingly large for them all to be isotopes of 'transuranic' elements of higher atomic number than uranium itself. Finally in 1939, the answer emerged, through a fortunate stroke of nuclear serendipity.

Hahn and Strassman were checking to see whether certain of the activities could be due to isotopes of radium, produced perhaps by rapid and repeated decay, through emission of two alpha particles, of the irradiated uranium. This unlikely explanation (since no such alpha emissions were in fact detectable) seemed to be supported by the finding that the radioactive material was 'co-precipitated' from solution by barium, a conveniently available reagent which was used because it was somewhat similar to radium in its chemical properties. The precipitation of barium salts from a solution should thus cause precipitation also of those of radium.

To clinch the matter, the final and careful precipitations were carried out with solutions of radium itself, to establish that the new radionuclides were indeed forms of radium. This, however, they obstinately refused to be; they continued to behave like barium, but not like radium.

Bohr heard from Frisch and Meitner that barium, which has atomic number 56, had been positively identified in this way as being formed by the neutron irradiation of uranium, of atomic number 92, and that this clearly suggested splitting, or fission, of the nucleus rather than progressive radioactive decays. Bohr's announcement of this finding, at a session of the American Philosophical Society in Washington, effectively broke up the meeting, with participants leaving hurriedly for their telephones or their laboratories. Four communications appeared in the *Physical Review* three weeks later confirming the neutron fission of uranium into smaller atomic fragments. Another report from Copenhagen to the same effect appeared in *Nature* in the same week, and one had been published from Paris in *Comptes Rendus*. The multiplicity of radioactive candidates for transuranic status was explained: they were different fission products of much lower, and not higher, atomic number and mass than uranium. The characteristics of nuclear fission, once recognized, were rapidly established by a combination of experimental and theoretical methods.

It was, for example, predicted from considerations of nuclear energy, and confirmed experimentally, that the nucleus would split into two unequal fragments, having atomic numbers approximately two-fifths or three-fifths of the original value; that fission would be associated with considerable immediate releases of energy and of radiations of various kinds; and that the likelihood that fission would be induced depended critically upon the energy of the neutrons used to induce it. It was shown, in studies that must have been somewhat inspired by hindsight, that fission was unlikely in elements with atomic number much below 100; and that, even in elements of high atomic number, the energy delivered by a neutron might not be sufficient to deform the nucleus enough to cause fission, or might be dissipated in causing other forms of radioactive emission.

In one of the isotopes of uranium, however, uranium-235, the 'excitation energy' delivered to the nucleus by the capture of a neutron of the most appropriate energy exceeded by about 20 per cent that needed to distort and split the uranium nucleus. In the resulting fission, additional neutrons were released, since the uranium nucleus contains more neutrons than do the fission products that are produced. In a block of uranium, some of these released neutrons will escape from the surface of the block, or be captured by non-fissile forms of uranium, or by any substances present as impurities. Some, however, will survive to cause fissions of other uranium atoms. If at least one neutron released per fission survives in this way to cause a further fission, and is at, or is brought to, the low energy that allows it to do so, there will be a self-sustaining chain reaction of fissions throughout the block of uranium. These conditions require a certain 'critical mass' of fissile material to be used, since the loss of neutrons from the surface of smaller masses will be too large to allow the reaction to proceed.

In naturally occurring uranium, only about 0.7 per cent is in the form of the uranium-235 which captures and is fissioned by low energy neutrons in this way. Over 99 per cent is present as uranium-238 which captures such neutrons but is not fissioned by them. Instead, the incorporation of the neutron produces the isotope of higher mass but equal charge, uranium-239 (Fig. 3.1). This decays by loss of an electron, so producing a nuclide of higher (positive) charge and atomic number but equal mass, neptunium-239. This in turn decays, again by electron emission, producing a fissile isotope of the element of next higher atomic number, plutonium-239.

Some knowledge of the effects of neutrons on uranium is necessary, therefore, in understanding the types of radiation exposure that may result from the

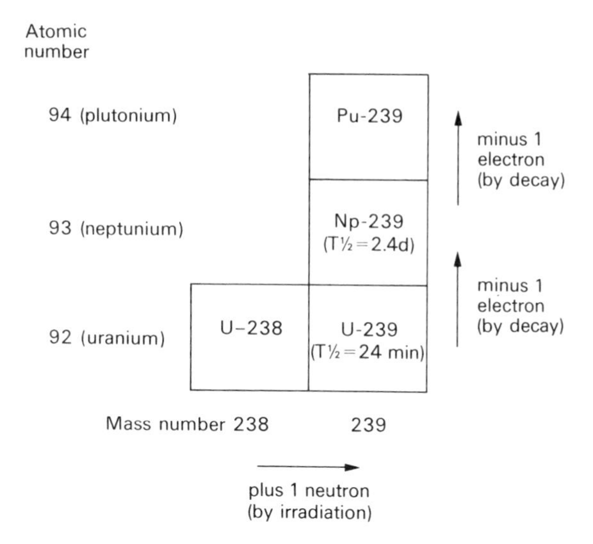

Fig. 3.1. Transformation of uranium-238 following neutron irradiation (to uranium-239, with successive decays to neptunium-239 and to plutonium-239).

fission process, either when it occurs in atomic bombs of the types that were exploded in Japan, or when it is used as the basis for energy production in nuclear reactors. In the former case, fission occurs almost instantaneously throughout a mass of uranium or plutonium, with unrestricted release of radiation and fission products. In the latter, the aim is to obtain a gradual and controlled rate of fission and energy production, with efficient containment of the fission products formed and of the radiations emitted during fission. The biological problems are however related in each case to the penetrating, gamma and neutron, radiations released at the time of fission and to the subsequent behaviour of the radioactive fission products; and, in the case of atomic weapons, to the very large releases of energy occurring at the time of the explosions.

ATOMIC BOMBS

The spread of fission through a mass of fissile material in an atomic bomb needed to be very rapid, since otherwise the energy released by the earliest fissions would scatter the remainder of the material, and the 'critical mass' needed to sustain the continuing fission reaction would no longer be present. This rapid spread is only obtained if most of the neutrons released during fission are available to cause further fissions, and if relatively few are captured by other materials. For this reason, no explosive effect can occur with uranium unless the concentration of the fissile isotope uranium-235 is very greatly increased, and relatively little of the non-fissile uranium-238 component is present.

Because all isotopes of any element have the same chemical properties, the isotope separation of uranium-235 from natural uranium required for such 'enrichment' could only be achieved by physical methods, depending on the small differences in mass of the various isotopes, with consequent very small differences in their behaviour in diffusion, or under centrifugal or electromagnetic forces. Elaborate techniques were developed, therefore, for the progressive fractionation of the uranium isotopes, to separate uranium-235 in sufficient amounts, and in sufficiently pure form, from the natural uranium to provide the explosive effect of the bomb that was released over Hiroshima on 6 August 1945.[18]

The explosive charge for the bomb detonated over Nagasaki three days later was different, and its preparation depended on the operation of a reactor. It had been predicted, from theoretical knowledge of nuclear structure, that as a final result of neutron capture in uranium-238 a material (plutonium-239) would be formed which would be likely to have a long radioactive half-life but be fissile by slow neutrons. During 1942, therefore, after extensive investigations, a quantity of uranium was assembled under conditions of wartime secrecy, at the University of Chicago. Enough highly purified graphite was obtained to slow the neutrons released in fission to an energy which would maximize the likelihood of their capture in causing further fission. Using these materials, and the necessary equipment for control and surveillance of the fission process, the first man-made reactor was constructed – in a squash court under the West Stands at the Stagg Baseball Field. The plutonium that was finally generated in this reactor could, of course, be separated in pure form from the remaining uranium and fission products in the reactor core by simple chemical methods, with no need for the complexities of isotope separation.[18]

The devastating effects of the bombs depended essentially upon the energy and the radiations released at the moment of the explosion, causing immediate fires, destructive blast pressures, and extreme local radiation exposures. Since the bombs were detonated at a height of some 600 metres above the ground, only a relatively small proportion of the fission products were deposited on the ground near the 'ground zero' point below the site of detonation. Some deposition occurred however in areas near to each city, owing to local rainfall occurring soon after the explosions. This happened at positions a few kilometres to the east of Nagasaki, and in areas to the west and north-west of Hiroshima.[19, 20] For the most part, however, these fission products were carried high into the upper atmosphere by the heat generated in the explosion itself. On their subsequent return to earth they contributed to human irradiation, although to an extent which is very small compared with that from the vastly greater activities discharged by subsequent atmospheric tests which returned to earth world-wide in fallout.

In Hiroshima, of a resident civilian population of 250 000 it was estimated that 45 000 died on the first day and a further 19 000 during the subsequent four months. In Nagasaki, out of a population of 174 000, 22 000 died on the first day and another 17 000 within four months. Unrecorded deaths of military personnel and foreign workers may have added considerably to these figures.[21]

It is impossible to estimate the proportion of these 103 000 deaths, or of the further deaths in military personnel, which were due to radiation exposure rather than to the very high temperatures and blast pressures caused by the explosions. From the estimated radiation levels, however, it is clear that the radiation alone would have been enough to have caused death within the first few days or weeks in the majority of those exposed within a kilometre of the ground zero below the bombs.

To these 103 000 deaths have since been added those due to radiation induced cancers and leukaemia, which had amounted to between 400 and 500 within 30 years of the explosions, and which may ultimately reach about 1000.[22]

The major source of exposure in both cities was from the penetrating gamma radiations, and to a lesser extent from the neutrons, emitted during and shortly after fission. To these exposures, two further, and smaller, sources were added. One, already mentioned, was due to the 'black rain' which fell in some areas, carrying down radioactive materials from within the rising cloud of fission products. The exposures due to these depositions are in general estimated to have been small, but some increased activity from the radionuclide caesium-137 remained detectable for many years in soil and farm products in the Nishiyama district east of Nagasaki.[20]

The second additional form of exposure resulted from the effect of neutrons in inducing radioactivity in various stable chemical elements, in the way that has already been discussed. The half-lives of decay of most of the radionuclides so formed are short, and the activities induced were not high, although high enough to allow the neutron exposure at ground level to be estimated approximately from the activity induced in iron or concrete structures or in roofing tiles. The total absorbed doses of radiation from these activities are estimated to be less than 1 per cent of that from the neutrons which induced them. They could however have caused a significant exposure of people who entered the city within a few days of the explosions.

WEAPON TESTS

The atmospheric testing of nuclear weapons caused people to be exposed to radiation in a quite different way. The atomic bombs had caused lethal exposures locally from radiation at the time of the explosions, but very little radiation at distances of more than a few kilometres from the ground zeros of the bombs. Subsequent atmospheric tests are not known to have caused any substantial exposures of people at the time of the tests, or from any transient radioactivity induced at the test sites. The fission products released into the atmosphere, however, caused the whole world population to be exposed to very low but continuing annual doses from fallout;[23] and, in two instances, caused substantial irradiation to small populations exposed to local fallout close to the site of testing, as described later (p. 77).

This is not a handbook of nuclear weapons. As a basis, however, for our assessment of the amounts of human exposure that have resulted, and will result, from past atmospheric testing, it is necessary to look briefly at the types of weapon tested, the forms and amounts of fission products produced, and the ways in which they enter what is rather ungeometrically called the biosphere, and cause irradiation of man and other species.

The 'atomic' bombs used in Japan, and the bombs or devices tested during the following seven years, depended on the fission of uranium-235 or plutonium-239. The explosive effect of each was equal to that of up to a few tens of thousand tons of the conventional explosive TNT (trinitrotoluene). On this basis of comparison, the Hiroshima bomb was probably of about 15 kilotons – that is, of 15 thousand tons of TNT equivalent – and that at Nagasaki was of 25 kilotons. In addition, the total equivalent of all atmospheric weapon tests made by the end of 1951 was in the region of 600 kilotons.

After 1951, however, devices were being tested which had explosive effects about a thousand times greater, and by the end of 1962 the total of all atmospheric tests had risen from the 1951 value of 0.6 million tons equivalent, to about 500 million tons equivalent.[23]

This vast increase in scale was due to the testing of 'thermonuclear' weapons, which depended, not on the fission of a critical mass of fissile material alone, but on a three-stage process initiated by this reaction. An initial fission, such as occurred in the 'atomic' bomb, momentarily created conditions of enormously high temperature and atomic disturbance that allowed the fusion together of the nuclei of atoms of low atomic number, such as lithium and hydrogen. This fusion not only liberated further large amounts of energy explosively, such as occurs in the similar reactions in the sun and stars. It also released neutrons of sufficiently great energy to set off the third stage of this fission-fusion-fission process. This third stage consisted of the fission of a surrounding 'blanket' of uranium, since even the uranium-238 isotope is fissionable by neutrons of this high energy. This third stage then caused a final and very large release of additional energy.

The yield of fission products is not necessarily proportional to the explosive power, but in fact is approximately so since much of the energy of the thermonuclear weapon is derived from fission processes. From 1952 to 1962 therefore, the amounts of fission products discharged into the atmosphere were of the order of a thousand times greater than all discharged previously.

To complete this tally of the total fallout to date, all atmospheric tests since 1962 appear to have increased by rather less than 20 per cent the total of fission products that had been deposited by previous tests, as judged by the measured deposition of strontium-90 in successive years.[23]

MECHANISM OF FALLOUT

The fraction of all fission products which is carried up into the higher levels of the atmosphere depends in part on the height above the ground at which the bomb is detonated, and in part on the power of the explosion, since it is the heat generated by the bomb which carries most of the cloud of fission products rapidly into the stratosphere.

Local fallout

Fission products may however be deposited near the site of the explosion, whatever its power, if the detonation occurs on, or at a small height above, the ground surface, or at a shallow depth below it so that cratering occurs. Such local deposition is increased if coral or rock components of the ground are vaporized, so that dust particles are formed which are large enough to fall back to earth rapidly. These mechanisms are likely to have caused much of the local contamination of atolls and lagoons near the testing sites at Bikini.

In addition, however, local fallout will occur if rain falls through the cloud of fission products as it rises and spreads downwind. Such fallout occurred on to several inhabited islands in the Rongelap and Utirik atolls, and on to the crew of a Japanese fishing vessel which was within 100 miles of Bikini at the time of a thermonuclear test in March 1954. The effects of the resultant exposures, principally on the thyroid gland in the first case and on the skin in the second, are discussed later (p. 144).

World-wide fallout

By far the greatest part of the fission products that have been released by atmospheric tests have been carried up into the stratosphere at heights above 15 km. The air in this region mixes only slowly, and at particular latitudes and times of year, with that of the lower atmosphere. No precipitation occurs directly from within the stratosphere, in which the air is extremely dry. It is found, in consequence, that fission products that have been carried to this altitude only return slowly to the lower, tropospheric, atmosphere, after an average time of about two years, by which time components with short half-lives will largely have decayed to non-radioactive nuclides. Once in the troposphere, however, the remaining activities are relatively quickly deposited onto the earth or sea, either by rainfall or by the mechanism known as dry deposition.[14]

During the period spent in the stratosphere, longitudinal transfer and mixing occurs rapidly by jetstream action, but latitudinal transfer across the equator is slow. Each hemisphere thus receives mainly the fallout that is initiated in it.

Once deposited at the surface, the components of fallout circulate in ways which depend on the chemical nature of the radionuclides involved, and on the types of surface on which they fall. These 'pathways' have been closely studied, in their connection both with fallout and latterly with nuclear wastes. In particular, the rates and extent to which different radioelements move from water or land surfaces into materials that may be eaten as food, or into the air that is breathed, are reasonably well established under a variety of conditions. In addition the average amounts of external radiation that will be received from a surface or subsurface contamination of soil can be estimated for different gamma emitting radionuclides, allowing for shielding by ground cover, housing materials, and other factors. All this allows a confident, if approximate, estimate to be made of the average radiation doses received from the longer-lived fallout components which return from the stratosphere, as well as from the short-lived nuclides which are present also in 'fresh' fallout deposited locally. The sizes of the resultant radiation doses are discussed in Chapter 6, in relation to those from other sources of radiation exposure.

REACTORS AND THE NUCLEAR FUEL CYCLE

The reactor which went into operation in the Chicago squash court in December 1942 not only produced plutonium for military purposes, but also produced evidence that power could be generated under controlled conditions by a self-sustaining chain reaction of uranium fission. The types of reactor which have been developed in the subsequent 40 years have differed in the ways in which the heat is removed from the reactor core, in the 'moderating' materials used to reduce the energy of the neutrons released in fission so that they are likely to cause further fissions, in the extent to which the uranium of the reactor fuel is 'enriched' by increased concentrations of the fissile uranium-235, and in numerous other respects.[24]

From the point of view of radiation exposures, however, we are concerned essentially with the efficiency and reliability with which the radiations emitted as a result of fission or of induced radioactivity are contained within the reactor, and with the exposures that may be caused by any discharges of radioactive materials from it.

Exposures occurring in these ways from reactor operation, however, form only a part, and normally only a minor part, of the total exposure of people attributable to the whole nuclear fuel cycle on which the operation of the reactor depends – a cycle which extends from the original mining of the necessary uranium ore to the ultimate disposal of wastes and the dismantling of old reactors when they are finally decommissioned.[25] Some outline needs to be given, therefore, of the steps involved in this cycle, and the ways in which people may be irradiated at each stage, as a basis for estimating the amounts of radiation, and therefore the amounts of harm, that are likely to be involved in producing energy from nuclear sources.

Uranium mining and the preparation of nuclear fuels

The isotopes of uranium decay through a long series of radioactive daughter products before isotopes of lead with a stable nuclear structure are finally reached. Uranium ores, and the rock from which they are mined, therefore contain a variety of gamma emitting, as well as alpha emitting, radionuclides. Uranium miners are consequently found to receive quite substantial amounts of gamma radiation during work underground, although surface mining involves much lower exposures.

The associated alpha radiations from the rock do not cause any significant external radiation exposure, since their penetration through body tissue is so very limited. This penetration may however be just sufficient to pass through the dead cells on the surface of the skin and reach live cells below it, and some excess of skin cancers on the face in Czechoslovak miners has been attributed to dust deposited on the skin.

The essential importance of alpha radiation in mining depends, however, not on external but on internal radiation from inhaled dusts and gases. Part way down the chain of daughter products radium is generated, and this in turn decays slowly to the radioactive gas radon. When radon is inhaled, its own daughter products, which include alpha emitting forms of polonium, lead, and bismuth, become deposited on the walls of the air passages. This causes lung and bronchial irradiation which has in the past given rise to very high death rates in miners. (Indeed, the mortality in the early pitchblende miners used to be so high that Agricola writing in 1556 alleged that some miners' wives were known to have had seven husbands, in succession.[26]) An increased death rate from lung cancer is still reported in uranium and other hard rock underground miners in a number of countries, despite greatly improved ventilation and working conditions, and is mainly attributable to the inhalation of radon.

The problems of occupational exposure are similar in type but substantially less in magnitude during the subsequent stages of preparing the uranium for use as a nuclear fuel. After the ore has been milled, the uranium is extracted and purified. For most types of reactor, the concentration of uranium-235 is enriched by physical processes before the resulting oxide or metal is loaded into the rods in which it will be used in the reactor core. During these stages, the uranium is separated chemically from its radium and other daughter products, so that the subsequent occupational exposures both to gamma radiation and to radon are progressively reduced. Some public exposures may result from these operations, however, from releases of radon or other decay products of uranium from the 'tailings' of waste material that are deposited near uranium mines or mills. The size of such exposures depends on the distance from centres of population of such mines or mills, which is normally large, and also depends critically on the methods used for covering and containing tailings which are no longer in use (as discussed on p. 159).

Nuclear power reactors

In the reactor, the fuel rods are contained within a core which is surrounded by heavy shielding to minimize the escape of neutron, gamma, and other radiations emitted during fission. This shielding also reduces the radiation from the fission products which accumulate as the fissile material in the rods is used up, and from radioactive nuclides formed in reactor components by the effects of neutron irradiation. Some external radiation exposure of reactor operators results from these sources nevertheless. This may arise particularly during maintenance work when, although the reactor is shut down and no fission is occurring, some radiation from fission products and from induced radioactivity is still likely to penetrate parts of the reactor shield (p. 151).

The stream of circulating material, whether liquid or gaseous, which is used to extract heat from the reactor core is likely to contain radionuclides of two kinds: fission products which have escaped from within the fuel rods through any defects in their cladding; and any material in which radioactivity has been induced by the high neutron flux within the core. As the coolant material issues from the core, the greater part of most of these radionuclides is removed by various filtration or other treatments. Some activity, however, is liable to escape into the air or water environment or to be released as part of waste management procedures, and may give rise to human exposure.

Under accident conditions much greater releases could result, particularly under conditions in which the fuel rods become overheated so that the integrity of their cladding becomes locally defective and fission products escape into the stream of coolant and thence into the environment (p. 155).

Our estimates of all human irradiation resulting from reactor operation must therefore take account of the amount of occupational exposure from external and internal radiation, and exposures of the general public from radioactive materials that may be discharged from the reactor site, both in the course of normal operation and in the event of accident.

Reprocessing of nuclear fuel

In the reactor, the fission products are largely retained within the fuel rods in which they are formed. In reprocessing, however, the spent fuel is removed from the rods, after they have been stored for periods of up to several years to allow the shorter-lived fission products to decay. The unused uranium and the generated plutonium are separated chemically from the accumulated fission products, the former components being prepared for reutilization and the latter for storage and later disposal.

Possibilities of human irradiation therefore arise at various stages, and total environmental exposures are ordinarily greater than those resulting from reactor operation.[23] On the other hand the possibilities of any large accidental releases are smaller, since no power production is involved, and both the chemical separation processes and the continuous cooling of highly radioactive fuel and wastes present fewer technical problems than the exact control of the fission process in a power-generating reactor.

The evaluation of occupational exposures must include external radiation received during storage of fuel rods or their contents before and after, as well as during, the chemical separation processes. They must take account of the possibilities of internal radiation from any materials escaping into the working environment, particularly once the fuel rods have been opened. Similarly, assessment of public exposures must involve atmospheric and aqueous discharges and disposal of solid low activity waste during any of these phases, and the routes by which they may return to man.

Waste disposal

The amounts of human exposure that may result from the ultimate disposal of high activity wastes are necessarily much more difficult to predict, in view of the very long half-lives of some of the alpha emitting radionuclides present in these wastes, and the consequent impossibility of any direct experimental study of the way in which such materials might migrate from their initial sites of disposal over time periods of tens of thousands of years.

THE OKLO REACTOR

Surprisingly, however, such an 'experiment' has already been performed for us, from which some indication can be obtained of the probable mobility of such alpha emitters, including plutonium, through rock formations over periods, not just of thousands, but of many hundreds of millions of years.[27, 28] It appears that the first nuclear reactor was not the one assembled in Chicago in 1942. The real precedence goes to a critical assembly of uranium which started operating in Africa some 1700 million years earlier, and is estimated to have generated a few hundred kilowatts of energy for several hundred thousand years.

This natural reactor developed within rock strata containing a high content of uranium ore. The present-day concentration of the fissile uranium-235 in natural uranium is only 0.7 per cent. This is insufficient to sustain a nuclear reaction without the use of special moderators to reduce the energy of the neutrons released in fission. Millions of years ago, however, the relative concentration of this fissile isotope will have been considerably higher, since uranium-235 decays rather faster (if a half-life of 700 million years can be described as fast) than the non-fissile uranium-238 (with its half-life of 4500 million years). At the time when the reactors at Oklo, in Gabon, were operating, therefore, the fissile uranium-235 was in a sufficiently enriched state to sustain a chain reaction, at least at times when water penetrated the strata and moderated the neutron energies, and until the heat generated had been sufficient to boil off the water and temporarily close down the reactor.

This digression into palaeoradiology is of significance, not only perhaps in confirming that nothing is ever really new, but for the indication that it gives that the plutonium that was formed appears not to have migrated through the strata in which it was formed, despite their permeation by water. The plutonium itself has long since decayed, since its half-life is (only) 24 000 years. The products of its radioactive decay, however, are limited in position to just those areas in which the fission reactions had occurred, these areas being identifiable today by the presence in them of abnormal concentrations of various nuclides resulting from previous fission. The plutonium, therefore, appears to have stayed for the whole period of its radioactive decay in the positions in which it was formed, and not to have migrated from these positions despite the permeation of the strata by water.

The careful, inch by inch studies of nuclide distribution in the Oklo rocks, therefore, give evidence that, at least in one set of conditions, plutonium does not migrate from the site of its deposition in rock strata. Current estimates of the possible exposures that might result from the disposal of high level wastes, however, are ordinarily based on the maximum estimated rates of transfer of long-lived radionuclides that could occur over long periods of time from an initial containment in sites of burial below the land or ocean floor.

4

MEASUREMENT OF RADIATION

The amount of harm that may be caused by radiation depends in a relatively simple way upon the quantity of energy from the radiation which is absorbed by the body, or by different parts of the body. To assess the importance of different sources of radiation in harming people, therefore, we must have information of three kinds.

First we must know how much radiation is delivered to people from all the various sources that have been described. We need to know the average exposures, as well as the highest exposures, of individuals from each source; we need the numbers of people exposed to each; and we need to know whether the exposures involve the body as a whole, or certain body organs or tissues only. The present chapter describes the methods of measuring or estimating these quantities of radiation. Chapters 5 and 6 then review the amounts of radiation that we are receiving from the different sources.

Secondly, we will need to look at the information required to link the size of these radiation doses with the types and frequencies of harmful effects that they may cause.[29] Subsequent chapters will therefore describe the way in which we have obtained our present evidence on the immediate and later effects of large radiation doses, the causation of malignant, hereditary, and other abnormalities by radiation, and the frequency with which these effects may be induced by different amounts of radiation exposure.

The third step can then be taken of reviewing the number of such effects that have been, or may be, caused by different radiation sources or practices.

RADIATION QUANTITIES AND UNITS

Quantities of energy are measured in 'joules'. The joule is defined on an electrical basis, but one joule is an amount of energy that would raise the temperature of one gram of water by about ¼°C or would lift a one kilogram weight through ten centimetres. We are interested in the amounts of energy delivered to different body tissues as a result of the ionizations caused by radiation in these tissues, since the harmful effects of radiation on body cells depend upon the number of these ionizations. A useful index of the likely harmful effect of a given radiation exposure is therefore obtained by measuring or estimating the number of joules of energy delivered by ionization, per kilogram of tissue.

Grays and sieverts

In the conventions of the SI system (the 'système international d'unités'), the unit of 'absorbed dose' of radiation in any material is the gray, one gray corresponding to the delivery of one joule per kilogram. (With the adoption of the SI system of units, the gray replaces the former unit of absorbed dose, the rad, one gray corresponding to 100 rads. L.H. Gray was a pioneer in radiobiology. There was no Dr Rad.)

This simple measure of energy delivered per unit mass is valuable for comparing the effectiveness of different amounts of radiation of a particular type, for example of gamma or of alpha radiation. When, however, we are assessing the effects on the body of a mixture of radiations of different types, as from exposure to gamma and alpha radiations together, it is a nuisance to express the total dose in grays, since each gray of alphas causes more harm than each gray of gammas. This is essentially because the individual ionizations are much more closely spaced along the short track of the alpha particle in tissue, than in the long track of gamma radiations. The likelihood of irreversible damage, for example by simultaneous hits on both DNA chains of a chromosome, is therefore greater per unit of energy delivered by alpha radiation that it is with gamma radiation, when a single hit caused by one track traversing one DNA chain of a chromosome has a substantial chance of being repaired before the other chain of the same chromosome is traversed by a separate, second track (as discussed on p. 96).

Whatever the mechanisms involved, it is preferable to use a unit of dose which allows for the observed differences in biological effectiveness per gray of different types of ionizing radiation. This more convenient unit is called the sievert (Rolf Sievert having laid the foundations of modern radiation physics). For most forms of radiation, including x rays, gamma, and beta radiation, the sievert simply corresponds to an energy deposition of one joule per kilogram, and so is equal in size to the gray. Alpha radiation, however, is substantially more damaging per gray than x rays and gamma radiation, by an amount which varies with the energy of the alpha particles and with the size of the doses delivered; and in many circumstances each gray of alpha radiation is taken to correspond to 20 sieverts. For neutrons also this 'relative biological effectiveness', or RBE, varies with neutron energy and other factors, but in general each gray from neutrons is conventionally counted as equivalent to 10 sieverts.

The 'dose equivalent', expressed in this way in sieverts, is therefore a better index of the total biological effectiveness of a given exposure to multiple types of radiation than is the purely physical measure of the absorbed dose of energy expressed in grays. The amounts of radiation received by members of the population from different sources are therefore given in the following chapters in sieverts (Sv); or more commonly and conveniently in millisieverts (mSv, each equalling one thousandth of a sievert), since the average dose received annually by the population from all natural sources is typically of about 2 millisieverts, with all artificial sources, including medical radiology, adding rather less than 1 mSv. (The sievert is the official SI unit which replaces the former unit of dose equivalent, the rem, with 1 Sv corresponding to 100 rems.)

Collective doses

Several other terms will be useful when we are considering the total impact of a particular radiation source upon the population.[30] It is then important to take account not only of the dose, or average dose, that people receive from a given source, but the number of people who receive these doses and so the total harm for which the source may be responsible. For this purpose it is useful to estimate the collective dose attributable to the source, in man-sieverts, indicating the number of people exposed multiplied by the average dose that they receive. Unless the probability of harm is strictly proportional to the dose received, however, over the whole range of doses involved, the collective dose will only give an approximate basis for estimating harm, since the harm caused when 100 people each receive one sievert may be greater than that from 100 000 people each receiving one millisievert.

Doses in the future

It will also be important to recognize that the intake into the body of a radionuclide which is retained, for example in bone or lung, for long periods may irradiate these organs for the rest of the lifetime of the individual. In these circumstances, the relevant dose is that to which he is 'committed', depending upon the retention characteristics of the nuclide and upon the age of the individual at the time of intake.

A similar 'dose commitment' into the future occurs when a particular discharge of radioactive material will continue to cause irradiation of many generations of people, so that the dose to the current generation is no longer an adequate measure of the radiation exposure due to the discharge. The estimation of a 'collective dose equivalent commitment' to future populations, however, becomes as complicated as its terminology, since it depends on guessing population sizes and practices far into the future.

Genetically significant dose

When hereditable abnormalities are induced in the germ cells of an individual by irradiation of his or her gonads, the likelihood that these abnormalities will in fact be inherited depends on the number of children that the individual will conceive subsequently. In Great Britain, the 'child expectancy' falls with age from an average of about 2 per individual at birth to values of less than 0.2 at ages of over 40 in males and 35 in females.

For a group of people who are all exposed at the same dose level, therefore, the number of induced abnormalities which are in fact inherited will depend on the age structure of the group, with fullest expression from those aged less than about 18, and with very little contribution from those aged over 40. The fraction of the collective dose to a population which is 'genetically significant' in this sense may thus vary considerably, and be twice as high in a staff of young radiographers as in one of ageing radiologists. In the population as a whole, the genetically significant fraction of a collective dose delivered to the gonads equally at all ages is about 0.4 (0.42 in males and 0.34 in females in England and Wales in 1977). In estimating the 'genetic' harm due to irradiation of a population, therefore, the risk that hereditable damage will be induced by the exposure needs to be related to the number of individuals at ages which will allow any such effects to be expressed.

Effective dose equivalent

In estimating the harm from different radiation exposures it will be necessary to distinguish situations in which the whole body is irradiated more or less uniformly, and those in which certain body organs only are irradiated, as commonly occurs when radionuclides have been inhaled or swallowed. Enough is now known of the 'sensitivity' of different organs to cancer induction or other damage by radiation to provide a rough scale of comparison of the probability of harm due, for example, to 1 Sv delivered to lung, thyroid or bone marrow, or to the same dose delivered to all body organs equally in whole body radiation exposures.[29]

This has value when assessing the radiation exposure from natural sources, when it would be misleading to add the number of millisieverts of whole body radiation from cosmic radiation, to the millisieverts delivered to bone alone from radium, or to lung alone from radon. A preferable alternative is to use the known estimates of the relative risk of irradiating lung or bone alone, compared to that from whole body irradiation; and to express the dose from all natural sources in terms of the whole body dose that would involve the same total risk of harm. The lung or bone doses are weighted according to the risks per millisievert which apply to them when irradiated alone.

This use of an 'effective' dose equivalent has two defects. It relies on numerical estimates of the risk of exposure of different organs which, as discussed later, are reasonably reliable for the six or so organs in which the risks appear to be highest, but which are very approximate in other organs for which the risks have been shown to be low or may be absent. And secondly, it involves arbitrary weighting factors to be given to the induction of a major hereditary defect by irradiation of the gonads, and the causation of a fatal cancer by irradiation of other tissues. It remains a convenient, if very rough, way of indicating the approximate probability of harm from a source causing various types of exposure,

Table 4.1
Estimated risk of irradiating single organs relative to that of the whole body uniformly

Organ or tissue	Average risk (male and female, per thousand per Sv)	Relative weighting factor
Risk of genetic defect		
Gonads	4.2	0.25
Risk of fatal cancer		
Bone marrow	2.0	0.12
Lung	2.0	0.12
Breast	2.5	0.15
Thyroid	0.5	0.03
Bone	0.5	0.03
Other	5.0	0.30
	16.7	1.00

The weighting factors are derived to apply to personal risks from occupational exposure if individual organs are selectively irradiated. The genetic risk is that of defects expressed in children or grand children. The fatal cancer risk to 'other' organs is that to the five unlisted organs which receive the highest doses, with a weighting factor of 0.06 to each.

and of comparing different sources in this way, when a detailed listing of the probable frequency of each type of harmful effect might make the intercomparison of sources harder rather than easier.

We shall need to return to a review of the types of harm resulting from irradiation of the different organs, and their relative frequencies following whole body exposures at low dose, in Chapter 11. As an example of the derivation of weighting factors used for estimating effective dose, however, Table 4.1 gives the values recommended by ICRP and widely adopted internationally, for the risks assumed to be involved in exposure of different organs, and the relative weighting factors for these organs that result.[30] These estimates give equal weight to the causation of any major inherited defect by genetically significant irradiation of the gonads, and to the induction of a fatal leukaemia or other cancer by irradiation of the bone marrow or other organ. The weighting factors are of use particularly in determining the annual occupational limits of intake of different radionuclides into the body, to ensure that a given total risk is not exceeded whatever the organs irradiated, and the amounts of their radiation, by each radionuclide. For this purpose, and where a simple guide is required as to the safety of some hundreds of different radionuclides, the risk estimates used here are average values for all adult ages and for both sexes, in the light of present available evidence.

METHODS OF ESTIMATING RADIATION DOSES

Our knowledge of the amount of radiation that people receive from different sources is based on a variety of methods, the choice of method depending upon the type and amount of exposure. In general, doses due to external exposure from sources outside the body are, in the case of those exposed to radiation at work, measured by dosimeters worn by the individual worker on his clothing. External exposures of the public, however, are usually too low to measure in this way and are estimated by using sensitive instruments recording radiation levels at fixed positions, by surveys with portable instruments, or by calculations based on knowledge of the environmental transfer of the different radionuclides.

The assessment of internal dose, on the other hand, depends primarily on estimates of the amounts of different radionuclides that enter the body, and knowledge of where each is retained in the body, and for how long. In many types of occupational exposure, and in a few forms of public exposure, direct measurements on the individual can be made to establish the internal dose that he is receiving. For most types of public exposure, however, the activities incorporated in the body are too small to be measured with any precision, and the doses from them are more reliably assessed in the light of the probable intakes and metabolic behaviour of the relevant radionuclides.

ESTIMATION OF EXTERNAL RADIATION DOSES

It is possible to measure directly the amount of energy delivered to water or other material by radiation, and so to determine the absorbed dose in joules per kilogram of the material exposed. This can be done by 'calorimetric' measurements of the heat produced in the material, with corrections for the small proportion of the energy which does not appear as heat.

These methods, however, are very insensitive at the low levels of radiation to which people are exposed: twice the annual dose received from all natural sources would be needed to raise the temperature of body tissues by one millionth of 1°C. Sensitive and accurate determinations can, however, be made of the number of ionizations produced by radiation exposures, using electrical or chemical methods. The size of the corresponding absorbed doses can then be reliably inferred from knowledge of the amount of energy required to produce each pair of ions.

In practice, therefore, simple methods of this sort can be used to determine the amounts of radiation which reach the body in different circumstances. Three devices are commonly used for measuring occupational exposures: the ionization chamber, the film badge, and the thermoluminescent dosimeter.

Ionization chambers

The ionization chamber measures the number of electrically charged ions produced, by allowing this ionization to take place in a volume of gas enclosed between electrically charged plates. As the ions caused by the radiation exposure are attracted to, and deposited on, these plates, the charge carried on the ions causes a current to flow through the instrument, and the amount of ionization can be estimated from the size of this current.

Simple ionization chambers are made that can be worn like a pen in the pocket. In this case, one of the charged 'plates' is a fine quartz fibre, of which the position can be read on a scale seen by looking down the barrel of the 'pen', through a lens at one end of it. As the ions caused by radiation exposure are deposited on the plates, the electrical repulsion between the plates is reduced, and the quartz fibre moves across the scale by an amount depending on the number of charged ions generated in the instrument since it was electrically charged before use. From the amount of ionization in the instrument (which is in effect a miniature electroscope) it is possible to infer the ionization that will have been induced in body tissues below the position at which the chamber was worn, and hence the dose to these tissues. The scale against which the position of the fibre is seen can therefore be calibrated in units of radiation dose, and offers an easily read warning of increasing exposure during any work in areas in which there may be a high external dose rate.

In occupational conditions in which exposure at high dose rate might occasionally occur, monitoring devices can be installed also at normal working positions, so that alarms will be sounded if such exposures are occurring. These devices may respond to an increased rate of ionization resulting either from raised radiation dose rates or from increases in radioactivity of air drawn through the sampling equipment.

Film badges

For most conditions of occupational exposure, in which the dose rate is unlikely to vary abruptly or unexpectedly, adequate surveillance is maintained by records of the dose received during weekly or monthly periods. For many years, these measurements have been based on the original observations of Röntgen and of Becquerel, that x rays or radiations from radioactive materials cause fogging of photographic plates or films. A small strip of film is worn on the surface of the working clothes, in a covering which excludes light but not radiation. After one or several weeks the film is developed, by normal photographic methods, and the degree of blackening compared (by the use of a photometer) with the blackening of a series of standard films, exposed to known radiation doses as measured by ionization chamber or other methods.

Estimations of ionization and hence dose by the use of photographic film also offers a convenient way of checking the efficiency of shielding of radiation sources, or the contribution to occupation exposure due to time spent in various working positions. Exposures from natural sources are usually more simply and accurately assessed by ionization chamber, but they have been usefully examined by using personal badge type dosimeters in studies of the dose received in an area of high natural radiation background in Southern India. Here increased dose rates result from the thorium content of a coastal strip of monazite sands on which the houses of a fishing community are built, the houses offering only a limited shielding from this source of radiation.

Thermoluminescent dosimeters

External radiation doses can also be measured by so-called solid state detectors. In certain materials, when exposed to radiation, electrons are displaced from their normal positions in the crystal structure of the material, the number displaced varying with the amounts of radiation to which they were exposed. The displaced electrons return to their normal positions when the material is heated, and light is emitted as a result of their change in energy state when they do return in this way. Measurement of the light emitted by the thermoluminescent dosimeter, or TLD, can therefore be related to the energy delivered by the radiation to which it has been exposed, and thus to the dose received by the body tissues underlying the position at which the TLD was worn.[31]

Both with the film badge and with the TLD, the measured dose will only estimate the whole body dose if the radiation exposure is uniform over the body surface. In situations in which the exposure is known or suspected to be non-uniform, the use of the TLD offers the advantage that measurements of dose can be obtained with very small areas or amounts of thermoluminescent material. Thus if work involves the hands being much closer than the body to a source, a small sheet of TL material can be worn on the finger to ascertain the amounts of exposure received by the fingers.

Both with film badges and with TLDs it is usual to design the holder so that different parts of the sensitive film are shielded with different types or thicknesses of covering (Fig. 4.1). The changes registered below these various 'windows' then gives an indication of the penetration, and so the energy or type, of the radiations responsible for the exposure, and thus may help to identify its cause as well as its importance.

ESTIMATION OF INTERNAL RADIATION DOSES

The dose delivered to a body organ by any radionuclide that is retained in it depends upon the concentration of the radionuclide in the organ, the length of time that it remains there, the number of atoms that undergo radioactive decay during this time, and the energy released during each such disintegration. Information is also needed on the fraction of this energy which is absorbed within the organ, and the energy delivered by 'crossfire' from any other organs containing the radionuclide.

Fig. 4.1. Thermoluminescent film holder showing 'windows' in the holder (open circular window admitting beta and gamma radiation, and window shielded by domed polypropylene cover admitting gamma but not beta radiation) allowing separate measurement of highly penetrating and less penetrating radiations. (Reproduced by courtesy of the National Radiological Protection Board.)

This is less difficult than it sounds, as a great deal of information has been obtained on the way in which most of the important radionuclides become concentrated and retained in different organs or parts of the body after they have been inhaled or swallowed. There is information also on the extent to which they will be absorbed into the body from stomach, gut, or lung, according to the chemical form of the radionuclide, or its physical state and particularly its solubility, when it was inhaled or swallowed. With knowledge also of the types and energies of the radiations emitted by each radionuclide and so the penetration of these radiations through body tissues, it is usually possible to form an approximate but reasonably reliable estimate of the likely doses that will be delivered to different body organs when a given amount of a radionuclide is taken into the body in a particular form and by a particular route.[32]

In situations in which radionuclides may enter the body, therefore, the problem is either to estimate directly what radionuclides and what amounts of each have entered the body; or to estimate what types and amounts are likely to be entering, from measurements of their concentration in air, food, or water. The methods that can be used for these purposes depend upon which radionuclides may be involved and the amounts in which they are, or are likely to be, present. For most radionuclides, direct measurements can be made, either by whole body counters or by excretion studies, on those who are occupationally exposed to them, to confirm that the body content of these radionuclides, or the intake of them, is sufficiently far below the occupational limit of exposure to them. In the case of members of the public, incorporated activities are ordinarily too low to be detectable by such methods, although examples have occurred in which incorporation from world-wide fallout (p. 56) or from releases from nuclear plants (p. 178) were measurable. Reliance can, however, be placed on knowledge of the amounts of radioactivity discharged into, or present in, the environment, and on the extensive studies that have been made on the exposure that may result from particular levels and types of environmental contamination of the air, the sea or rivers, and the soil or foods, under different conditions of water usage or dietary practice.[33, 34]

Whole body counters

For those radionuclides which emit any penetrating gamma radiation during their decay, the amounts present in the body can commonly be estimated with considerable sensitivity by the use of counters outside, or surrounding, the body to measure these radiations. Each time an atom in the body decays in this way, a 'photon' – the type of particle of which gamma radiation is composed – is emitted. If this photon emerges from the body and passes through one of the counters, the ionization caused in the counter by its passage is detected electrically. The number of such passages, the count, is recorded. It is then a relatively simple matter to link the total count rate recorded by an array of counters surrounding the body with the frequency of decays, and so the radioactivity, within the body, by calibration methods using a model of the body (oddly termed a 'phantom') containing known amounts of different radionuclides.

Clearly the efficiency of a whole body counter depends upon the number, size, positioning, and sensitivity of individual counters used, and the extent to which radiation coming from sources other than those within the body can be excluded. For this reason the counter array is ordinarily set up in a room with heavy steel walls and door, often using steel plates from battleships built in the days before constituents of more modern steel were liable to be contaminated by fallout or other sources of radioactivity (Fig. 4.2). Similarly, extraneous radioactivity is minimized by a change of clothing prior to counting, and by cleanliness and 'barrier' procedures to prevent contamination being carried into the counting space.

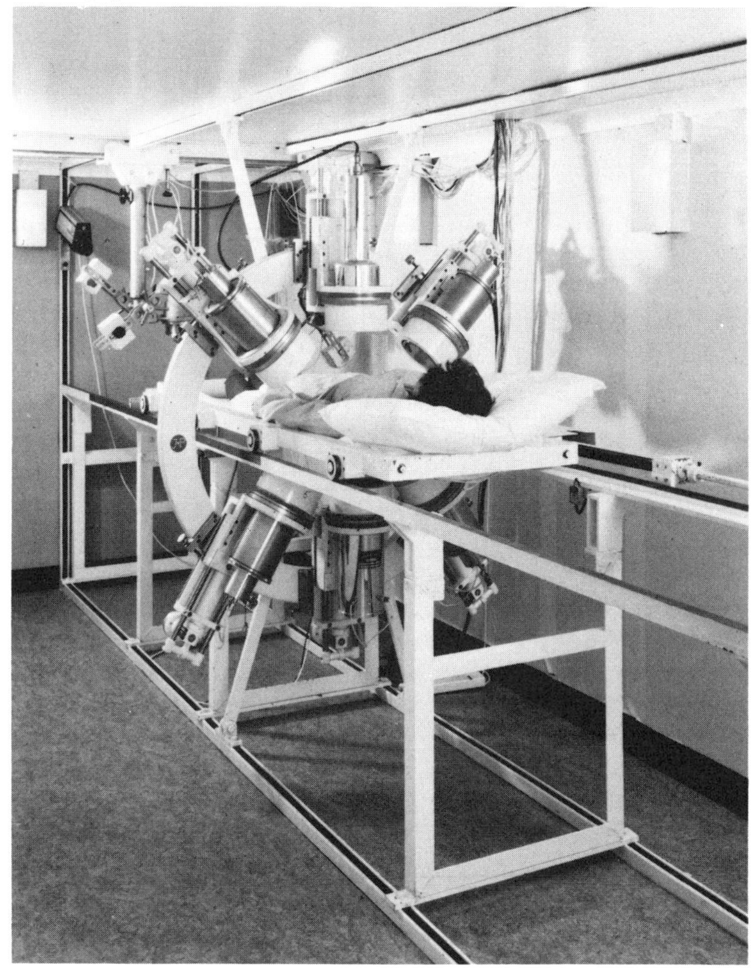

Fig. 4.2. Whole body counter assembly and array of counters for measuring body radioactivity. (Reproduced by courtesy of the National Radiological Protection Board.)

The counters that are ordinarily used depend on the effect of radiation in displacing atoms from their regular positions in a crystal structure. They have the important property that they respond not only to the number of photons that pass through them, but also to the energy of each photon. Since different radionuclides emit photons of different characteristic energies, the record of the number of photons of different energies which pass through the counter per unit time allows an estimate, not merely of the total gamma emitting radioactivity of the body, but of the radionuclides to which this is mainly due. The electronic techniques of 'curve stripping' allow the contribution of major components to be subtracted, so that that of minor contributors, or of radionuclides of particular importance, can be investigated.

The capacity of these crystal counters to distinguish the body content of different radionuclides is of obvious importance in identifying the likely cause of any significant occupational contamination. At lower levels of activity, they distinguish the contribution due to the natural levels of body radioactivity, chiefly from radium in bone and from potassium-40 present in body tissues generally. The method is so sensitive that it was possible to make accurate measurements of the body content of radioactive caesium (caesium-134 and 137) in members of the general public during periods of high fallout in the 1960s, or in individuals after eating sea fish containing these radionuclides owing to discharges from a local nuclear fuel reprocessing plant.

An alternative counting method, of even higher sensitivity but somewhat lower power of discriminating between different radionuclides, is of value in medical procedures, when the body content of potassium, or of a small dose of administered radionuclides, requires to be measured or followed. This method, which depends on measuring the scintillations produced in certain liquids by radiation, allows the body to be much more completely (and far more cheaply) surrounded by tanks containing these liquids than by any practicable array of crystal counters, so that photons from a greater proportion of the atomic disintegrations pass through the sensitive volumes of the counters, and are recorded.

For radionuclides emitting only radiations of weaker penetration through body tissues, special external counting methods may be needed. For example, the measurement of any plutonium-239 present in the chest depends on detecting the small proportion of the weak x rays which are emitted following its decay, and which pass through the chest wall. In the monitoring of any workers who might have inhaled plutonium, the assessment of a possible lung contamination not only involves an accurate correction for counts arising from any other radioactivity in the rest of the body and in the environment. It also necessitates a direct estimate of the thickness of the chest wall (conveniently by ultrasonic methods) since this critically affects the fraction of these weak radiations which escape from the body in each individual.

Measurement of excretion rates

The rate of excretion of a radionuclide in the urine can often be related more or less accurately to the amount present in the body. The regular measurement of urinary radioactivity, and identification of any radionuclides present, is therefore a simple method for detecting the occurrence, type, and amount of occupational exposure to many radionuclides. It is of particular importance when the radiations involved are too weak to be emitted from the body surface. Samples of urine taken at intervals of several weeks will detect or exclude the occurrence of single accidental intakes if the excretion from the body of a single intake is sufficiently prolonged. For some elements which are rapidly excreted, routine monitoring at practicable intervals may underestimate or miss some intakes, although measurements after a suspected contamination will still be reliable in estimating its occurrence or size. In some cases, however, including that of inhaled materials of low solubility, the progressive migration within the body over long periods of time, for example from lung to lymph glands and to bone or liver, excludes any unique or simple relationship between the urinary excretion rate and the radioactive content of any body organ or of the body as a whole. In such cases, the relation between excretion rate and body content may vary with time since intake, and average values for this relationship need to be assessed empirically from detailed study of instances of known accidental contamination.

Other methods of monitoring

Monitoring by whole body or partial body counting and by urine examinations may be supplemented in certain circumstances by measurement of breath, nasal secretions, or faecal samples, or by direct assessments of the activity of any contaminated wounds. It is evident, however, that the evaluation of average levels of occupational exposure to internal radiation is less easy and less comprehensive than that for external radiation where, whatever its imprecisions, the constant use of some form of dosimeter at work provides a sound basis for estimating the approximate levels of external whole body exposure, year by year, in most circumstances.

It does appear, however, that occupational exposure to internal radiation is, on average, materially lower than that to external radiation, so that total 'effective' doses, and therefore the associated risk, can ordinarily be estimated with reasonable reliability.

This remains true in the two occupational situations in which internal dose constitutes a substantial proportion of the total annual dose. In uranium and other hard rock mining, the average annual dose resulting from inhaled radon usually exceeds that from external radiation. Here, however, estimates of the amount of radon inhaled have been made by means of personal air samplers carried by the worker himself. These measurements validate, or supply a correction to, the estimates based on samples taken in fixed positions in the mines. Also, where intake of tritium (hydrogen-3) gives rise to an appreciable fraction of total dose, particularly during work in heavy water moderated reactors, the excretion of tritium from the body is sufficiently slow for regular urinary sampling at several week intervals to give an adequate estimate of the internal dose from it.

Environmental models of radionuclide distribution

As regards the exposure of the public to internal radiation from environmental sources, the levels of exposure are ordinarily too low to be measured directly in individuals by the methods used in monitoring for occupational exposure. As already mentioned, radiocaesium has been detectable during the 1960s following its deposition in fallout, and in the 1970s after its release from a reprocessing plant, and radioiodine was measurable in urine of Marshall Islanders from fallout in 1954, and in the thyroid glands of a few people living near Windscale after releases from a reactor in 1957. In addition strontium-90 has been detectable in specimens of bone obtained at autopsies in many countries since the periods of substantial fallout, particularly in the 1960s.

In general, however, estimates of internal dose need to be based on environmental measurements of the concentrations of radionuclides in air, water or food materials, and on models relating these concentrations to the intake, or commonly the maximum likely intake, by any members of the public.

A great deal of work has been done, therefore, on factors determining the exposure of people, not only as exposed locally near any sites of release or deposition of radionuclides, but also when the exposures result from any nationwide or worldwide distribution of fallout materials or components of nuclear wastes. The study of atmospheric and ocean circulatory systems, and of the food chains by which discharged radionuclides may return to man, has been actively pursued in a variety of contexts since the mid-1950s, when the worldwide fallout from nuclear weapon tests and releases of fission products necessitated the evaluation of the exposures that were being received, which were mainly from internal radiation, in consequence of them. The United Nations radiation committee (UNSCEAR, see p. 162) reported extensive work in progress in a number of countries already in 1958 on the rates of transfer of various fission products through the atmosphere to the soil, into vegetation, and into milk and other sources of human diet.

In the subsequent 25 years, such work on environmental pathways and food chains, and on the numerical values of the relevant rates of transfer and transfer coefficients, has extended to a range of fission products and transuranic elements of significance in the nuclear fuel cycle. Thus the UNSCEAR in 1977 was able to derive realistic estimates of the size of the human whole body or organ doses likely to result from given amounts of over 50 different radionuclides, if released into different aqueous or atmospheric media.[29]

Work is continuing on such subjects as the resuspension into seawater of shallow coastal sediments, the long term circulation within the ocean of radionuclides with very long half-lives, or the variation between individuals of different ages in the metabolism of nuclides which are slowly transferred from one body 'compartment' to another. Enough is known, however, to make it unlikely that major pathways of the return of released radionuclides to man have been overlooked, and to allow at least approximate estimates to be made of the amounts of internal radiation that are being received from different sources by the population as a whole, and by the more highly exposed groups within the population.

COMPREHENSIVENESS OF DOSE ESTIMATES

It is evident that a variety of methods can be used to estimate the doses that people receive from natural sources of radiation, from medical and occupational exposures, and from other local or global sources of ionizing radiation. It appears that reliable estimates can now be made of the total amount of radiation that we are receiving from all sources, and the average amount from each. Moreover, the contributions from so many of these sources have been so fully examined in different countries and under different conditions that we can now assess, not only the typical or average exposures from particular sources, but approximately the extent by which these exposures vary in different countries and in different circumstances. The next two chapters will review the sizes of these exposures and of their variation.

5

HOW MUCH RADIATION DO WE RECEIVE? AND FROM WHAT? NATURAL AND MEDICAL SOURCES

In order to assess the amount of harm that people may be receiving from different sources of radiation exposure, we need to estimate the doses of radiation that they receive from each source, and to express these doses in units which, as far as possible, indicate the harmfulness of the exposures. As already discussed, the 'dose equivalent', in sieverts, estimates the energy delivered to body cells and, with its allowance for the differing effectiveness of different types of radiation, gives a useful indication of the probability that different forms of harm may result in the various body tissues.

This chapter deals with the major amounts of human exposure to radiation, namely those from natural sources and from medical procedures. Chapter 6 then estimates the further doses received from occupational exposures, from nuclear power production, and from other sources and practices. We can expect the annual doses quoted in these chapters to give a measure of the relative amounts of harm due to the different radiation sources present in the environment. When these doses are combined (in Chapter 11) with the risk of harm per unit dose, we can reach a numerical estimate of the probability of harm in an individual receiving such doses, or the number of harmful effects to be expected in a population exposed to a certain average dose from each source of radiation.

Some effects of radiation do not occur at all unless quite high doses, of several sieverts or tens of sieverts, are exceeded. These effects probably result mainly from the killing of considerable numbers of cells in certain body tissues by these high doses, as discussed in Chapter 7. Severe radiation sickness or death may follow exposure of the whole body to doses of two or more sieverts if delivered in a short time (of hours or minutes); or severe damage to various organs of the body may develop gradually, if these organs are selectively exposed to higher doses, usually of some tens of sieverts, delivered over periods of months or years.

No such late effects are likely, however, unless body or tissue doses exceed a few tenths of a sievert per year. In most cases, the annual doses to which people are currently exposed are much lower than this, whether from natural, occupational, or environmental sources of exposure or from all these together. Only under accident conditions, or following some forms of radiotherapy, are problems of radiation sickness or late impairment of organ function liable to arise.

For the most part, the average annual doses reviewed in this chapter are of a few millisieverts – thousandths of a sievert – or less. Their importance, therefore, is as a basis for estimating the number of cancers or inherited defects that may be being caused by exposure of people at these low dose rates, since these effects of radiation may be caused even by the lowest doses. The number of inherited abnormalities or cancers is not necessarily directly proportional to the size of the dose at all dose levels. In comparing the consequences of low doses from different sources however, the number of such effects which is attributable to each source will be about proportional to the dose delivered by each source. The annual dose rates reviewed in the following sections, and summarized in Tables 5.2 and 6.4 (pp. 73 and 91), therefore reflect the relative importance of cosmic radiation, medical radiology, continuing fallout, nuclear wastes, and other sources, in causing hereditary defects or cancers. In the case of hereditary defects, it should be remembered (p. 47) that it is the 'genetically significant' part of the population exposure that is relevant; that is, the doses delivered to those at ages at which they will still have children. For cancer induction, the 'effective dose' (p. 48) takes proper account also of the likelihood of cancer induction when different body tissues are unequally exposed.

In these chapters we will need to be concerned not only with the average levels of exposure throughout a population from different sources, and the variations in these levels in different individuals, but also with the variation in these levels in different countries or parts of the world. In this, the detailed and documented studies of the United Nations Committee, UNSCEAR,[23, 29] are of particular value. The bases on which these doses are estimated are described in some detail, since readers will wish to judge with what confidence the different contributions to our total radiation exposure can be assessed.

EXPOSURE FROM NATURAL SOURCES

The human race has always been exposed to natural sources of radiation in three ways. To these we have added a fourth by living in houses (or caves). These types of natural exposure result from cosmic radiation, from the normal radioactivity of our body tissues, from radionuclides in the soil or underlying rock, and from the radon which accumulates in our houses, and in the less draughty kinds of cave. The exposures from the first three of these sources are usually about equal. That from the fourth is commonly rather larger.

Doses from cosmic radiation

The earth's upper atmosphere is constantly being bombarded by high-energy particles, which originate mainly from within our own galaxy. This 'primary galactic radiation' consists of a mixed bag of particles. Mainly these are protons, of which the energy – as measured by instruments carried on balloons or spacecraft – varies over the enormous range of some 14 orders of magnitude (i.e. 14

factors of ten). Helium ions also contribute, with a few heavier atomic nuclei which were discharged from exploding supernovas; smaller numbers of electrons and other subatomic particles are also present. As judged by the proportions of the different atomic nuclei, the particles of this galactic radiation have been on their way here for a few million years.

At times of solar flares, the galactic radiation is accompanied by protons and alpha particles emitted from the sun, but these are mainly of low energy, and do not penetrate the atmosphere to reach the earth's suface.

When the particles of higher energy start colliding with the atoms of the earth's atmosphere, they react to produce cascades of particles of lower energy, in a manner analogous, even if not very closely analogous, to the opening phase of a game of snooker. Particles of this 'secondary cosmic radiation' reach ground level largely in the form of neutrons, electrons, and muons (subatomic ionizing particles derived from the decay of pions; if that helps). The annual dose from cosmic radiation is about 0.30 millisieverts, to which neutrons contribute only 0.02 mSv.

Variations in the cosmic ray dose rate

This value of 0.3 mSv per year applies to regions at latitudes greater than about 50°. Nearer the equator, the earth's magnetic field prevents some of the particles from reaching the earth, so that the annual dose at the equator is about 0.2 mSv.

At increasing altitudes, however, the variation is much greater, since there is then a thinner layer of atmosphere to reduce the intensity of the radiation. At the top of Mount Everest, for example, the rate would be in the region of 20 mSv per year. At somewhat more habitable altitudes, however, say at 12 000 ft., the rate would be 2 mSv per year.

There is some variation also during the period of the sun's 11-year cycle of activity, and the dose rates quoted above are average values. These average values, however, appear to have remained the same – to within 10 per cent – for half a million years, as judged by the amount of radioactivity induced by cosmic radiation in meteorites (the age of the meteorites being estimated by the decay of the primordial radionuclides which they contain).

Doses from body radioactivity

All body cells contain potassium, and a small percentage of this potassium – about 0.01 per cent of it – is in a radioactive form. This potassium-40 isotope has a half-life of decay of rather over a thousand million years, and is one of the primordial radionuclides which were incorporated in the earth when it was formed. It causes an almost uniform radiation of the whole body, with a dose of 0.17 mSv per year. This value is determined by measuring the potassium concentration in the tissues, knowing the fraction of all potassium which is radioactive, and knowing the proportion of these potassium-40 atoms which distintegrate per second (very few) and the energy delivered per disintegration.

Potassium is an essential constituent of body cells, and its concentration in the tissues is accurately controlled in health and even in most diseases. (Deposits of fat in the body contain fewer cells than most tissues, and so less potassium. Whether a person is fat or thin, however, his body organs get the same exposures from the potassium-40 in them.) Since the radioactive isotope is, of course, chemically indistinguishable from all other potassium, and is uniformly mixed with it, the dose rate from postassium-40 is the same in people on any diet or in any country.

Much smaller doses of uniform whole body radiation are delivered from an even longer-lived form of the element rubidium (Rb-87, half life 5×10^{10} years) which is also stable enough to have survived since the earth's formation, and since the even earlier supernova explosion in which it was synthesized. Small doses are delivered also from 'cosmogenic' radionuclides which are formed by the action of cosmic rays upon stable elements in the earth's atmosphere. Certain radioactive isotopes of hydrogen, carbon, sodium, and beryllium are formed in this way, and have half-lives long enough (several thousand years in the case of carbon-14) for them to become distributed through the biosphere and through body tissues. They only add a dose of 0.01 mSv per year to the 0.17 mSv per year delivered by the potassium-40.

Certain tissues, and particularly the lung or bone cells, are also irradiated selectively by the primordial radionuclides uranium-238 and thorium-232 and their decay products which occur naturally in the environment. The largest doses from these materials are those from inhaled isotopes of radon and their decay products in houses. These doses depend on the construction and locality of the houses, and are discussed below (p. 67). Apart from radon inhalation, however, this group of naturally occurring radionuclides delivers an average effective dose of 0.16 mSv per year, largely from polonium-210 and radium-226 which are taken into the body in small quantities in the diet and become concentrated in bone.

The total effective dose rate, from incorporation of naturally occurring radionuclides into the body, is therefore 0.34 mSv per year.

Variations in internal radiation dose rate

As already mentioned, the annual contribution of 0.17 mSv from potassium-40 does not vary. The dose from polonium-210 varies a little with smoking habits, since inhalation of tobacco smoke may raise it somewhat (e.g. from 0.12 mSv to 0.18 mSv per year). It is however much more greatly increased, and by a factor of up to 10, under the exceptional conditions in the arctic in which the reindeer or the caribou form a large component of the diet. The winter food of these animals is largely of lichens which accumulate substantial amounts of polonium-210.

The dose of 0.04 mSv per year from the remaining radionuclides of the uranium and thorium series is locally increased in a few parts of the world where drinking water is drawn from uranium-rich rock strata, or where the soil content of uranium is high, as in parts of Southern India or Southern Brazil. Here the annual dose from these materials may reach 0.05 or 0.06 mSv for small numbers of people. No large populations are likely, however, to receive annual doses much exceeding the average value of about 0.34 mSv from natural sources of internal radiation.

Doses from terrestrial radioactivity

It is estimated that some 95 per cent of the world's population receives an annual dose of external radiation of between 0.21 and 0.43 millisieverts from radioactivity in the soil or in the underlying rock. This dose is almost entirely due to gamma radiation from the Big Three primordial radionuclides, potassium-40, uranium-238, and thorium-232, or their decay products. Beta radiation from the ground contributes less than 0.01 mSv per year.

These values are based on measurements that have been made, usually at a height of one metre above the ground surface, and typically at some hundreds of locations in each of a number of countries.[23] These outdoor measurements vary somewhat with weather conditions, since soil moisture or snow cover give some shielding against the radiation.

Measurements of gamma radiation inside buildings may be lower or higher than those outdoors, according to whether the building materials are more effective in shielding against the environmental radiation, or in adding to it by their own radioactive content. The rate indoors ordinarily varies between about 0.8 and 2.0 times that outside, with ratios of 1.4 and 1.5 for typical brick and concrete buildings, and ratios below 1.0 when construction is mainly of wood. The estimated average value of 1.2 for houses as a whole, however, implies that the assessment of an average annual dose that people will receive does not depend at all critically upon the fraction of the day that they are assumed to spend within buildings.[23]

On the basis of surveys both indoors and outdoors in a number of countries and in different conditions, the average dose rate from terrestrial sources can be taken as 0.32 mSv per year.

Variations in dose rate from terrestrial sources

Individuals living in the same district will receive somewhat different exposures according to the construction of their houses. As already indicated, however, these variations are not large. The dose rates will not usually differ by more than a factor of two between the lowest and highest values in the same district, unless certain synthetic materials of substantial radioactivity are used in building construction. (The use of these synthetic materials affects also the exposure to radon, and they are discussed below, on p. 68.)

The variation in dose rate in different districts, however, is often as large as this and occasionally much larger. This variation depends upon the geological history of the region, and on the radioactivity of the material from which the rocks or soil were formed. The three important primordial radionuclides, potassium-40, uranium-238, and thorium-232, are present in higher concentration in the earth's crust than in the deeper layers of the planet's mantle or core. (They got there because their characteristics of solubility or density as oxides brought them towards the surface when the earth was still molten.) The concentration of these materials in sea water is low.

As a result, sedimentary rocks, which were formed originally as marine deposits, are in most cases of low radioactivity. Igneous rocks – those formed from the molten state – show higher radioactivity. Basaltic rocks originating in oceanic areas were formed of lava drawn from layers of the earth's mantle below the ocean bed, and are thus usually only moderately radioactive. Radioactivity is however considerably higher, typically by a factor of about 10, in igneous rocks formed in continental areas. These volcanoes yielded granites when the lava came from within the earth's crustal layer, or andesite rocks of rather lower radioactivity when the volcano's roots reached partly through the crust into the deeper mantle.

Radioactivity may thus be high locally in three circumstances:

1. Where rocks of igneous origin, and particularly granites, are exposed, or where the soil is derived from the weathering of such rocks.
2. Where materials released from such rocks have accumulated locally. This applies particularly to certain minerals such as monazite or zircon which occur as dense, mechanically resistant particles in highly insoluble form, and which may be washed ashore as coastal sands derived from the weathering of offshore rock.
3. In a few instances, where surface waters have arisen from radioactive layers of the crust, and contain soluble and long-lived decay products of the primordial radionuclides, of which radium-226 is the most important.

Rocks of igneous origin The first of these causes of variation in dose rate is much the commonest, and is responsible for one of the greatest variations. Quite large areas are known in a number of countries, and particularly in granite hill or mountain districts, where the gamma radiation dose rate is two or three times the average for the country. Although many of these districts are rather sparsely populated, the total number of people exposed in them is considerable. At some positions within these districts the dose rate exceeds ten times the average.

Perhaps the record, however, is held by a small hill in Southern Brazil, which was formed of granite from volcanoes which developed when Africa and South America still formed one land mass. On this hill, the Morro do Ferro, gamma radiation dose rates have been recorded which are over 500 times the normal rate.[35] The hill, however, is uninhabited except by scorpions and some rats, which do not seem to mind.

Monazite sands Stretches of monazite sands occur in several parts of the world. In the Espérito Santo and Rio de Janeiro provinces of Brazil, outdoor dose rate measurements have been made in and around three small towns built over such sands. Some of the highest readings were found in uninhabited areas. Those made in the towns, however, would indicate annual gamma radiation dose rates[35] in the range of 1.5 to 3.5 mSv per year to some 6000 people, and of 6 or 7 mSv per year to a further 6000.

In the Kerala and Tamil Nadu states of India, a population of 70 000 people live on a narrow coastal strip of sandy country, in which the monazite has the high thorium content of 8 to 10 per cent. Direct measurements have been made of the doses actually received by the inhabitants of this district,[35] a number of whom, for long periods and in the interests of science, wore the medallions of attractive design and religious significance, which incorporated thermoluminescent dosimeters (see p. 52). The results of this devotional exercise indicated an average dose received by the whole population of about 4 mSv per year, with 6 per cent of the population receiving over 9 mSv per year – or nearly 30 times the usual terrestrial dose rate.

Radioactive water sources Little information has been obtained on the external radiation doses that result in areas in which water emerges with high radioactivity. It is clear, however, that environmental radiation levels may sometimes be substantially raised for this reason, and dose rates of up to 10 mSv per year have been measured in the immediate vicinity of springs of mineral water of high radium content near Ramsar in Northern Iran.[35]

Summary of dose rates from natural sources, other than radon

The three natural sources of radiation described so far, therefore, cause an average total effective dose of about 1 millisievert per year (Table 5.1). Small communities may receive 2 or 3 times this total dose if they live at high altitude, or depend largely on reindeer or caribou in their diets; or may receive 5 to 10 times the average dose, if living in certain areas of granite or monazite formation. (Happily, nobody living high on a granite mountain is known to subsist on caribou.)

Table 5.1
Radiation from three natural sources

Source		Average annual dose	Major sources of increase
Cosmic radiation		0.30 mSv	to 2 mSv at high inhabited altitudes
Internal radiation		0.34 mSv	to 1.5 mSv on some arctic diets
Terrestrial radiation		0.32 mSv	to 10 mSv in some small communities
	about	1 mSv	

The annual doses shown in Table 5.1 express also the rate of genetically significant irradiation of the population, since the three natural sources specified give about equal exposures at all ages and to all parts of the body. They therefore estimate the dose rate delivered to the germinal tissues of those who will subsequently have children, as well as measuring the average dose rate to the whole body at all ages.

Exposures from radon[23]

The fourth type of natural, or at least habitual, exposure results from inhalation of two of the isotopes of radon. Both are present at concentrations in the air of houses which are commonly 10 to 20 times those in the air outside. Both are gases which are formed mainly by the decay of radionuclides present in the soil or in building materials. Both decay in turn to radionuclides which deposit on the linings of the air passages and in the lung.

Radon-222 is formed by the decay of the long-lived radium-226. It has a half-life of about 4 days. It decays through isotopes of polonium, lead, and bismuth successively. These decay products continue to irradiate lung tissues for some years before finally decaying to a stable form of lead.

Radon-220 is formed similarly by the decay of longer-lived nuclides. It has a much shorter half-life of about one minute, decaying also to stable lead through a series of radionuclides of short half-life. The total dose resulting from inhalation of radon-220 is about one quarter of that from radon-222. (The amounts of stable lead accumulating in the lungs from the decay of radon are entirely trivial compared with amounts accumulated following inhalation of stable lead from other environmental sources.)

Most of the radon-222 in the air of houses has been formed during the decay of radium-226 present in building materials and in the soil beneath the house. The contribution from the latter depends upon the radioactivity of the local soil and on the efficiency with which the living-space of the house is sealed off against entry of gases released from the soil. The amount of radon coming from the structure of the house depends critically upon the building materials and the rate at which the gas can diffuse through them. Smaller amounts of radon-222 are released into the air from the water used in the house, and from the use of any natural or liquified petroleum gas as fuel.

For radon-220 the only significant releases are from constructional materials. These, however, differ as widely in their content of the thorium-232 series of nuclides from which radon-220 is released, as they do in the uranium-238 series which releases radon-222. The very short half-life of radon-220 does mean, however, that even such coverings as wallpaper which delay diffusion of gas from the walls to the room, may reduce the air concentrations of this radionuclide considerably.

Sources of variation of dose rates from radon

Effectively, therefore, the main factors influencing the dose rate from inhaled radon are:

1. The radioactivity of the soil, as already discussed; or in a few instances, that of any materials used for 'filling' of reclaimed land.
2. The rate of ventilation of the building. Since the outside air is so much lower in radon concentration than that indoors, an increased number of air changes per hour will rapidly dilute the indoor radon with that outdated commodity, fresh air.
3. The type of building material.

The rate at which radon will enter the air of rooms from different building materials depends partly on their porosity and covering, and partly upon how they are used in the structure of walls, ceilings, or floors, but mainly on their radioactivity. In grading them for radioactivity, their content of radium-226, the long-lived progenitor of radon-222, is rather more important than their content of thorium-232, the even longer-lived head of the radioactive family which leads to radon-220. In fact, however, the concentration of these two index substances are sufficiently nearly equal in most building materials for the sum of their concentrations to give a good guide as to the range of variation to be expected in the amounts of radon that will be released from each material. (Phosphogypsum and alum shale concretes are exceptions to this, being high in radium content but not in thorium.)

Such a criterion can in any case only be approximate, since different kinds of brick, concrete, cement, stone, or other materials themselves each vary considerably in their composition. It is, however, easy to understand the very large differences in the radon concentrations actually observed in houses, when the differences in building materials are looked at in a quantitative way: for example, as their typical concentrations of radium-226 plus thorium-232.

On this criterion, there is a central group of materials of moderate activity, with average concentrations of 50 to 100. (The concentrations are quoted in becquerels per kilogram of material, one becquerel of any radionuclide being the quantity, or 'activity' of that radionuclide in which one atom undergoes radioactive decay per second.) This group includes many forms of brick, concrete, and cement. Lower values are found for natural gypsum, limestone bricks and many sands and gravels which have values of around 20; and wood is considerably lower.

On the other hand, rather higher values have been found for granite (about 200) and for various forms of tuff used in building (250 and 450). High values are also found in various by-products which are occasionally or locally used as building materials or incorporated in them. These materials include phosphogypsum, blast furnace slag, fly ash, alum shales, and bauxite brick. In products made from these materials, activities of 300 to 800 becquerels per kilogram have been measured.

In the light of such a wide range of activities, it is understandable that very large differences are observed between the air concentrations of radon in different houses. Surveys have been carried out in a number of countries to assess the average exposure from radon and its range of variation, as well as the way in which the radon concentration varies with time of day, time of year, barometric pressure, the use of sealants on walls or flooring, and the introduction into the house of radon in water and fuel supplies.

Average values for the total annual dose from all sources of radon in the home in 8 countries have ranged from 0.6 to 1.6 mSv per year (estimated as the effective dose due to the selective lung irradiation resulting from inhaled radon – p. 41). A considerably higher estimate of 3.7 mSv per year for Sweden is thought to be due to several special factors: the common use in building until 1975 of an aerated concrete containing alum shale; a relatively high release of radon from soil into houses in many districts of volcanic origin; an increasing use of stone in newer buildings; and the efficient air-tightness of houses allowing a decreased ventilation rate during the winter months in the interests of energy conservation – an illustration of the principle that You Can't Win.

For the other six countries in Europe and the two in North America for which estimates have been made, the average effective dose rate has been 1.0 mSv per year, but with very wide variations around this mean value in individual houses.[23]

Summary of dose rate from all natural sources

If this value of 1 mSv per year is included with that from the three other sources already reviewed, the effective dose from all natural sources averages about 2 mSv per year, half of this being from radon in buildings. The component from inhaled radon, however, delivers very little radiation to the germinal tissues, so the average genetically significant dose to the population from natural sources averages only 1 millisievert per year.

EXPOSURE FROM MEDICAL SOURCES

In assessing the extent to which the population is exposed to radiation for medical reasons, we have several problems.

First, why do it? The radiation involved is not a general environmental constituent in the way that it is when it results from natural sources, from fallout, or from the discharge of nuclear wastes. Only those people are irradiated who have x rays, 'isotope' tests, or radiotherapy, and to express their doses as an average dose per head of the whole population is a rather artificial thing to do.

Moreover, and following from this, to discuss the differences in annual dose in different individuals seems silly. Either you have had a few tens of sieverts to part of the body in treatment, or a few tens of millisieverts for diagnosis, or you haven't. There is not the continuous range of variation in dose to individuals that applies for other sources.

In addition, even the frequency with which diagnostic or radiotherapy procedures are used varies very widely in different countries, according more to the facilities available than to the medical needs of people. We can only look for illustrations, rather than for realistic world averages, of the contribution from medical radiology to the total irradiation of people from all sources.

And yet there do appear to be good reasons for assessing the amounts of radiation that are delivered for medical purposes – and reasons apart from a merely obsessive wish not to leave anything out of the balance sheet. One reason is that the amount of inherited harm from radiation looks at the total exposure of the germ cells of the population, or at least of the younger members of the population, and not at the reasons for the radiation exposure. We need therefore to include the genetically significant part of the dose from medical exposures alongside such doses from other sources, and see whether radiology is responsible for a large or small contribution to the total.

In much the same way, if all radiation exposures, however small, may contribute to the prevalence of cancer, it is surely relevant to include those doses from medical radiology which are delivered to body tissues at ages such that cancer might be induced by them; and to do so even though the values of the medical procedures are likely to exceed greatly the risks that may be involved in their use.

Doses from diagnostic x rays

Here we need to assess the doses due to diagnostic radiology as a whole, and the method of assessing them. The contributions from different types of x-ray examination need not be discussed in any detail.

In principle the method is simple. A record is obtained of the number of x rays of different kinds which are taken during certain periods at each of a number of hospitals, clinics, or other places where x rays are used diagnostically. The periods of time are arranged to cover different seasons, as the frequency of some x rays varies with time of year. The length of these periods, and the number of establishments included, are made large enough to assess adequately the frequency even of unusual types of x ray. The establishments from which records are obtained are chosen so as to sample properly the range of special hospitals, mass radiography units, or dental surgeries in the country or district studied. The ages of the patients examined are recorded, so that the genetic, and perhaps the carcinogenic, risk of the radiations can be taken into account. The radiation doses involved in each type of examination are assessed, including that to the germinal tissues and ideally those to the main body organs which are irradiated. This assessment may be done as a separate study, but should include at least some evaluation of the doses actually used in different hospitals or by different practitioners making the same type of examination and using the same number of films, as these doses may vary considerably.[36]

Finally the number of examinations and the total doses involved are scaled up to apply to all x-ray facilities in the district surveyed, and related to the population of that district.

This procedure has been carried out, in greater or less detail, in 19 countries and in 17 districts within countries, with repeated surveys in 13 countries.[29] In addition, information has been obtained from a number of other countries of the annual frequency per person of diagnostic x rays, of whatever kind. This latter information at least gives some indication of how much lower the average dose rate is likely to be in rural areas of countries with limited medical and transport facilities, than in the more industrialized countries from which most of the detailed dose estimates have been obtained.

The national and regional estimates take account of the proportion of diagnostic procedures which depend upon x-ray film; those which require, or still use, fluoroscopic examinations, with or without image intensification techniques which reduce the otherwise higher doses involved in fluroscopy; those by mass screening; and examinations by CT (computerized tomographic) scanning in which, as compared with simple x rays, a moderate increase in dose yields a great increase in diagnostically important information.

Estimates of the average genetically significant dose in 13 European and North American countries have averaged 0.3 mSv per year, individual national values ranging from 0.1 to 0.6 mSv per year.

These estimates, however, apply to countries with an average of about 0.75 diagnostic examinations per person per year. They must, therefore, grossly overestimate the genetically significant dose rate in many other countries for which direct estimates have not been made. The World Health Organization has published estimates of the frequency of diagnostic examinations in 28 countries of Africa, Asia, Oceania, South or Central America, and the Caribbean.[37] Of these, the annual frequency of examinations was less than 0.1 per person in 18 countries, suggesting that there the average genetically significant doses are commonly below 0.04 mSv per year per member of the population.

Similar large differences must apply to average whole body or effective doses to national populations. The doses delivered to various body organs during diagnostic examinations vary with the type and technique of the examination, but ordinarily range up to a few millisieverts, or sometimes tens of millisieverts, per examination. Taking account of the frequency of different kinds of examination, the effective dose varies from about 0.3 to 1.3 mSv per year in different countries; and the average value for a number of industrialized countries is about 0.6 mSv per year. This value is substantially higher than the genetically significant dose rate, because it includes a preponderance of examinations made at ages at which there is on average only a small likelihood of conceiving children subsequently.

Doses from diagnostic tests in 'nuclear medicine'

The method of estimating the average doses due to the diagnostic use of radionuclides (or radiopharmaceuticals) is similar to that used for diagnostic x rays. The frequency with which different procedures are used is recorded in a sample group of the various types of departments which employ these methods, and related similarly to the age of the patients and the use of nuclear medicine in the country as a whole. The only difference in method is that the dose delivered to the germinal tissues, and to the other body organs, is estimated from knowledge of the amounts, the distribution, and the retention in the body tissues of the radionuclides used in these tests, rather than by the organs exposed to the x rays and the number of x-ray films taken in other radiological examinations. Broadly speaking, nuclear medicine tests and diagnostic x-ray examinations give about equal genetically significant doses per examination; and the range of organ doses, and the average effective doses, are also similar in the two cases.

Few detailed reviews have been made recently of the average doses received from diagnostic nuclear medicine. Data are, however, available of the number of such investigations made per year and per person in several countries; of the number of the different types of test performed; and of the usual amount (activity) of the radionuclides used in each.[29] Since the dose to different body organs depends less on the technique of the test than in the case of x rays, a reasonable estimate is possible of the average annual doses in a number of countries, and this is supplemented by more direct estimates in a few.

Detailed studies in 1977 in Japan,[38] where nuclear medicine is widely used, indicated an average genetically significant dose from these procedures of about 0.004 mSv per year, and an effective dose of 0.02 mSv per year. These doses are 2.5 per cent and 1.5 per cent of the corresponding doses from diagnostic x rays in Japan (in 1979).

The frequency and type of radionuclide tests in four other industrialized countries also indicate that these tests add little, and probably less than 10 per cent, to the average genetic and effective dose rates from diagnostic x rays in countries with extensive medical facilities.[23] The same is likely to apply to countries with limited facilities, in which the frequency of nuclear medical tests is commonly lower by a factor of 10 or more than in some industrialized countries.[37]

Doses from radiotherapy

The doses delivered to individual organs during radiotherapy are considerably higher than those resulting from diagnostic procedures. Particularly during the treatment of cancer, whether by external radiation or in some cases by the use of radionuclides, these doses may reach several tens of sieverts in organs or tissues which need to be irradiated. The resulting radiation exposure, however, when expressed as an average dose per member of the population, is substantially lower than that from diagnostic radiology. This applies particularly to the average genetically significant dose to the population, since the great majority of treatments, and particularly those for cancer, are given at a time of life when further conception of children is unlikely. Moreover, any cancers which might be induced, and which might have developed only after the usual latency of 10, 20, or more years, are less likely to occur during the lifetime of the irradiated person. Even the risks per unit dose are reduced, because a proportion of the damaged cells which might have given rise to cancer are themselves killed by the high tissue doses needed for effective radiotherapy of existing disease.

The methods of estimating the average population doses due to radiotherapy, either by external radiation or by the radionuclides which are needed in certain thyroid or other diseases, are essentially the same as those used for diagnostic procedures. For countries with extensive medical facilities, the estimated genetically significant average dose from all forms of radiotherapy ranges from less than 0.005 to 0.025 mSv per year.[23] This wide variation probably reflects in part the extent to which radiotherapy is used to treat non-malignant disease, and in part the extent to which it is given in place of, or in addition to, surgery in the treatment of various forms of cancer.

The average effective dose has, as already mentioned, a reduced significance as an indicator of potential harm at the dose levels used in radiotherapy, and it has not often been assessed on a national scale. Judging from estimates of the total doses delivered to parts of the body or to the bone marrow, however, its value is probably in the region of 0.2 mSv per year. It is evident, however, that these estimates of mean population exposure would be considerably lower, and commonly by a factor of 10 or more, in many countries with limited medical and transport facilities.

Table 5.2

Annual UK exposures from natural sources of radiation and from medical radiology

Source of exposure	Usual range of doses, to tissues principally exposed	Average population dose rates, mSv per year: Genetically significant	Average population dose rates, mSv per year: Effective
Natural sources	1 mSv per year (body) +8 mSv per year (lung)	1.0	2.0
Radiology			
Diagnostic x rays	0.05 to 5 mSv	0.3	0.6
Diagnostic nuclear medicine	0.1 to 5 mSv	0.05	0.1
Radiotherapy	3000 to 60 000 mSv	0.025	0.2

Summary of doses from medical radiology

From the estimates already quoted, it is clear that the population exposure from diagnostic radiology exceeds that from radiotherapy. In industrialized countries, the average doses from medical radiology as a whole are about half those from natural sources (Table 5.2).

6

EXPOSURE FROM OTHER MAN-MADE SOURCES

Four other types of radiation exposure at present add about 0.03 mSv to the average annual effective doses of 2 mSv from natural sources and about 1 mSv for medical ones, as assessed in the last chapter. These remaining types of exposure may be grouped as those due to irradiation from fallout, from occupational exposures, from the use of nuclear power, and from various miscellaneous devices and practices.

DOSES FROM FALLOUT[23]

The peak years of atmospheric testing of nuclear weapons and devices were from 1954 to 1962. Tests during these years were responsible for most – about 85 per cent – of the nuclear fission involved in the whole sequence of atmospheric tests conducted so far (until the end of 1981). These years included an even greater proportion – about 95 per cent – of the total yield from fusion.

The peak year for global fallout was 1963, since 1961 and 1962 had seen by far the heaviest testing, with over 60 per cent of the total energy release occurring in these years, and since the radioactive products from large tests remain for some two years in the stratosphere before they return to earth.

Since then the scale of atmospheric testing, and the amounts of radioactive material falling out annually, have progressively decreased. These materials are now coming down at about 1 per cent of the rate recorded in the worst year – as indexed by measurements of strontium-90 deposition.

The present annual dose from fallout, of about 0.01 mSv, therefore gives only a partial and a rather confused indication of the radiation impact of atmospheric testing. This is partly because the present annual dose rate includes that fraction of the radiation from past fallout which continues to reach us from long-lived radionuclides; and partly because it does not take account of the doses that the human race will receive in the future from this year's fallout.

A fairer picture is given by the total dose that people have received in the past, and will receive in the future, from all radioactive products of all tests that have been carried out hitherto – provided, of course, that it is borne in mind that this is not an annual dose, but a total from all tests.

Doses from global fallout

This estimate can be reliably based on the large number of studies that have been made world-wide, of the amounts of radioactive material collected year by year at stations in many countries.[23, 39] These studies assess the proportion of different radionuclides in the deposited material, the redistribution that each of the main constituents will have after their entry into soil, air, water, fish, or animals, and so the extent to which people will be exposed to external radiation by the contamination of their environment, or to internal radiation by breathing air, eating food, or drinking water containing any of these radionuclides.

The average dose – past, present, and future – is estimated in this way to be about 4 mSv per member of the world's present population. The dose is somewhat higher, on average, for people living in the Northern Hemisphere than for those in the South – with 4.5 mSv in the North and 3.1 mSv in the South.[23] This is because more, and bigger, tests were carried out in the North, and the air in the upper atmosphere mixes only rather slowly across the line of the equator. In either case, the total dose commitment over all time from all tests to date is the equivalent of about 2 years' exposure from natural sources.

In fact this expression of an average dose per member of the population is misleading in one sense, since much of the total dose will be delivered far in the future, by carbon-14. This radionuclide is formed when neutrons, released during the fission or fusion processes in the explosions, are captured by nitrogen in the atmosphere. The carbon-14 has a long half-life (of 5730 years) and, mixing with other carbon in the biosphere, becomes distributed through body tissues and will cause irradiation at a very low dose rate, but over a long period of time. Very little of its dose will be delivered to the present generation, to whom the average dose from fallout will not exceed 1.5 mSv. Indeed, after the year 2000, no exposures from present fallout materials will be likely to exceed 0.1 mSv in a lifetime.

A transient increase in bone irradiation occurred in those who were children at the time of the heaviest fallout. Because of the chemical similarity between strontium and calcium, the radioactive strontium-90 in fallout became incorporated in newly formed bone. Young children in the late 1960s therefore had concentrations of strontium-90 in bone that were between 1.5 and 2 times the adult value. By the late 1970s however, these concentrations had fallen, owing to the turnover of bone mineral during growth, and its replacement by bone formed at times of lower dietary contamination.

A larger variation in dose from global fallout affected the arctic and subarctic communities already mentioned, since the reindeer and caribou on which their diet depended had a high intake of caesium-137 that had been deposited on the lichen which formed their winter diet.[40]

Doses from local fallout

The doses that have been received from local fallout contrast sharply with those from global fallout. In the latter, the radioactive products of atmospheric tests have remained in the stratosphere for several years, and the shorter-lived fission products have largely decayed away. Doses from global fallout have therefore rarely exceeded a few millisieverts, although delivered to very large populations.

In fallout occurring locally, near and soon after a nuclear explosion, however, small populations have been exposed to doses of up to a few sieverts. This is partly because short-lived fission products are still present, and partly because the deposition will have come from a much more concentrated cloud of nuclear debris.

In the areas in which some local fallout occurred in Japan, external radiation is estimated to have caused doses of from 0.1 to 0.5 sieverts near Hiroshima,[19] and of from 0.3 to 1.3 Sv in the Nishiyama district of Nagasaki.[20] Internal radiation also occurred, but is unlikely to have added substantially to these doses.

Two groups of people received substantial doses from local fallout in the Pacific, after a thermonuclear weapon test at Bikini in 1954.[41] The exposures in the Marshall Islands involved external radiation doses of up to 2 Sv to 82 inhabitants of Rongelap Island, and 0.15 Sv to 150 others on Utirik Island. Some 28 US service personnel also were exposed on a neighbouring island.

In addition, owing to the presence of short-lived radioisotopes of iodine in the fallout, the inhabitants of Rongelap will have received internal radiation to their thyroid glands, in doses that are likely to have been of about 1.5 Sv in adults, and probably between 4 and 14 Sv in children. The doses were higher in children because in them the iodine which is needed for synthesis of the thyroid hormone is collected into a smaller gland than in the adult. The concentration of radioactive iodine, per gram of gland, is therefore also higher. (The effects of these doses are described on p. 144.)

The crew of a Japanese fishing vessel was also exposed to fallout from this test.[42] The average external radiation to the skin of 24 members of the crew has been estimated to have been between 2 and 6 Sv. Thyroid exposures may have been similar to those of the adults on the islands, or lower since the latter resulted largely from drinking water collected from the roofs of buildings.

It is likely also that local fallout will have occurred near, and downwind of, some weapon testing sites, either following atmospheric tests of devices of lower yield than those mentioned above, or after any venting to the surface of the products of underground tests. No reliable data are available, however, as to the numbers of people who may have been exposed in this way or the doses that they might have received.

DOSES FROM OCCUPATIONAL EXPOSURE TO RADIATION

A great deal is now known about the doses that people receive in the course of work which may involve exposure to radiation, whether in medical radiology, nuclear plants, industrial radiography, or other occupations.[23] This is because regular measurements are made, usually at weekly or monthly intervals, on most people working in these fields, to assess the total amount of external radiation that they have received during the period. In those occupations in which radionuclides are liable to be taken into the body in significant amounts, tests are also made on a regular basis to assess doses received by internal radiation, by methods already described (p. 52). In such cases, however, a comprehensive estimation of dose is less easy than for external radiation.

In reviewing the exposure involved in different occupations, we should be concerned with three estimates:

1. The average annual dose received by people engaged in each occupation. In most occupations this average dose is of a few millisieverts per year. In a few, however, for example in uranium mining and in some activities involved in reprocessing or reactor maintenance, the average dose may lie between 10 and 30 mSv per year.

2. The frequency with which individuals exceed the average by any considerable amount in any one year; or, for example, the frequency with which any annual dose exceeds 15 or 50 mSv.

3. The contribution that each particular occupation makes to the radiation exposure of the population as a whole. This is indicated by the collective dose, in man-sieverts, obtained by multiplying the number of workers in the occupation by the mean dose that they receive. Its genetic significance of course depends upon the ages at which the exposures are received.

Doses to staff in medical radiology[23]

The radiation exposure of radiological staff has been assessed in a number of European countries, in the USA, in Canada, and in Australia. In diagnostic radiology and in nuclear medicine, the average annual doses ordinarily lie between 0.5 and 2 mSv. To dentists and their assistants the doses are lower – typically from 0.1 to 0.5 mSv per year. In radiotherapy the value depends somewhat upon the type of work and of worker: using treatment with x rays, cobalt sources, radionuclide administrations, or high energy machines; exposures of radiologists, gynaecologists, radiographers, or nurses. Average annual doses usually range from 1 to 2.5 mSv.

Large year to year variations are not common in any of these groups, and doses rarely exceed 15 mSv in any individual in any year.

The contribution that such occupational exposures make to the irradiation of the community as a whole depends predominantly on the radiological resources of the country. In 8 countries with extensive resources, the average collective dose was 1.0 man-sieverts per year per million of population, about half of this representing a genetically significant part of the dose. Approximate estimates in several countries with more limited medical facilities indicated collective doses of 0.2 man-sieverts per year or less, per million people. Compared with the annual collective dose from natural sources, therefore, of 2000 man-sieverts per million, these contributions to population exposure are very small.

Nuclear power production industries[23]

In many of the activities involved in the production of nuclear power, the average annual doses are in the same range as they are in medical work. Thus the average doses to staff are ordinarily in the range of 1 to 3 mSv per year in the operations involved in fabrication of the radioactive fuel elements, in enrichment of uranium, in various types of research and development, and in the operation of some forms of reactor. Rather higher rates, usually in the range of 4 to 12 mSv per year, may result from work on reprocessing, and in some aspects of reactor operation including the maintenance of reactor equipment. In uranium mining, and in a few of the activities involved in fuel reprocessing, higher rates of up to about 30 mSv per year commonly occur.

Despite the high average dose rates seen in a few of these occupations, individual doses greater than 40 or 50 mSv are rarely exceeded in any year, except perhaps in some uranium mines.[43] In underground mining the combination of internal radiation from inhaled radon and external radiation from the rock face gives doses which vary in different working positions and are difficult to assess with accuracy in different individuals. It is likely that an annual dose limit of 50 mSv may be exceeded during work in some conditions of poor ventilation or inadequate water screening of working faces.

Very much higher exposures have occurred accidentally on a few occasions during research on experimental forms of reactor, or when other assemblies of fissile material were brought together in such a way that they became 'critical' and an unintended fission reactor was started. A total of 10 such accidents have been reported in which exposures were sufficiently high to cause severe local injury or symptoms of radiation sickness, the last such recorded accident having been in 1965. Death occurred in 8 of the total of 38 workers exposed in these accidents, and doses of one sievert or more are likely to have been received by most of those exposed.[23]

In addition, numerous accidents of lesser severity have occurred, in which no clinical evidence of any radiation injury was detectable, but in which accidental exposure was indicated by an increased dosimeter record, or by evidence of radionuclide intake into the body by inhalation, by swallowing, or through a wound. The exposures resulting from such lesser accidents are in general recorded or estimated, and included in the assessment of the annual dose received from other occupational exposures.

The amount of occupational exposure from nuclear power production in any country can best be expressed in relation to the amount of electricity produced from nuclear sources. The total occupational exposure from all phases of the nuclear power cycle, including mining, reprocessing, and research, is estimated to deliver a collective effective dose of about 25 man-sieverts per gigawatt year's output of electricity from nuclear sources.[23] For example, therefore, in a country with a relatively high electricity consumption of 1 kilowatt per head, and deriving 20 per cent of this supply from nuclear sources (i.e. with 0.2 gigawatt nuclear supply per million people), these occupational exposures would give rise to a collective dose of 5 man-sieverts per million people. If expressed as an average dose rate per head of population, therefore, the contribution to the irradiation of the whole population from this type of occupational exposure would be 0.005 mSv per year. At this level of reliance on nuclear power the resultant occupational exposures would thus increase the average effective dose to the population from natural sources, of two mSv per year, by 0.25 per cent.

Other forms of occupational exposure[23]

For most other forms of occupational exposure, the average dose rates are lower. In a variety of types of industrial or other research, annual doses average 1 mSv or less, although a few such occupations using radioactive sources have involved rates of over 5 mSv per year. Aircrews may receive doses of up to about 2 mSv per year from the increased cosmic radiation at flying altitudes. Workers in metal mines may in some cases receive doses from radon which are as high as those in uranium mines, but ordinarily they are substantially lower; and they are lower still in most coal mines.

Radon is responsible also for high doses to the staff or attendants at certain 'radon spas'. The clients of one such spa in Austria are conveyed on a miniature railway through rock tunnels in which the air and water concentrations of radon are unusually high.[44] Whatever benefit the clients may derive from the experience, the engine driver has been estimated to have received an annual dose in the region of 200 mSv.

In two other occupations, however, larger numbers of people receive substantial doses. Tritium, the radioactive form of hydrogen, is now commonly used in place of radium to luminize watches and so to reduce the exposure of the wearer. The luminizers of the watches, however, have under certain circumstances received annual doses of 10 to 15 mSv of internal radiation from incorporation of tritium into the body, and doses of over 35 mSv occur in a few per cent of individuals each year. Substantial irradiation of bone used to occur also from incorporation of the radium used by the earlier generation of luminizers.

Significant annual doses may be received also from the use of radioactive sources for such industrial purposes as the radiography of metal castings and the logging of wells. The average exposures tend to be greater when portable sources rather than fixed installations are used. The use of such sources may involve regular annual doses of up to 5 mSv per year, with occasional substantial over-exposures, often from handling or use of the source without the appropriate shielding or equipment. Indeed, two deaths have occurred as a result of gross over-exposure while using industrial sources: in one case following a dose of 3.5 sieverts from a sealed radiographic source of iridium-192; in the other from a whole body exposure to 10 sieverts from a sealed source of cobalt-60 which was being used for agricultural purposes.[23]

Portable industrial sources involve also the severe added hazard that, if ever left unguarded and unsecured, they may be found and carried off by any child or other member of the public, and kept at home with no recognition of their dangerous nature. Such exposures are known to have occurred on at least 6 occasions during the last 20 years, and to have caused 8 deaths, either in children who found them, or in members of their families who were exposed to such unshielded sources in their homes (p. 99). It is a sobering thought that these distressing accidents have caused as many 'acute' deaths from radiation exposure as have resulted from all criticality accidents that occurred during the development of nuclear energy techniques.

Dose to the population from all occupational exposures

No useful figure can be given for the importance, world-wide, of occupational exposures in contributing to the genetic or the total irradiation of populations, since different countries differ so widely in their medical and industrial uses of radiation, and in their use of nuclear power.

Table 6.1

Genetic irradiation of the UK population due to occupational exposures, expressed as percentages of exposure from natural sources of radiation

Occupation	Contribution to population genetic dose, as a percentage of that from natural sources
Medical radiological staff	0.08
Nuclear fuel cycle (including reprocessing, 0.03)	0.06
Nuclear research	0.03
Industrial radiography	0.11
All other	0.15
	0.43

As an example, however, all forms of occupational exposure in the United Kingdom in 1978 increased by about 0.6 per cent the average effective dose received by the population from natural sources, and increased the corresponding genetic dose by 0.4 per cent.[45] The contributions from different occupations to this increase in population exposure were as shown in Table 6.1. This was under conditions in which about 12 per cent of electricity was being derived from nuclear sources. Fuel reprocessing was included within the fuel cycle, but no uranium mining or milling was involved.

POPULATION EXPOSURE FROM NUCLEAR POWER PRODUCTION

As in the case of occupational exposures, the amount of public irradiation from the nuclear power cycle needs to be related to the scale on which nuclear power is being used. It is therefore expressed in terms of the collective dose incurred per unit output of electricity derived from nuclear sources, so that the estimates may be applied to different countries at different periods of time.

Similarly, the total environmental impact can only properly be assessed when all stages of the cycle are taken into account, from the original mining of uranium to the final disposal of wastes. Admittedly, one country may use uranium which has been mined in another, or reprocess wastes which derive from another; and discharges of the longer-lived radionuclides into air and water in one country will contribute to the irradiation of people in other countries. A country may therefore receive more or less than the exposure corresponding to the full fuel cycle, according to whether it is, in effect, a net importer or exporter of man-sieverts from any nuclear power that is produced. The total collective dose per gigawatt year's output of electricity allows an overall assessment of the radiation impact of nuclear power production. It is however necessary to distinguish between exposures received locally near nuclear plants or regionally in the country of origin, and the more remote or global exposures due to the release and world-wide distribution of long-lived radionuclides.

Exposures of the public are essentially due to the radionuclides released in the course of nuclear operations, and hardly at all to any direct gamma or other irradiation from the fission processes going on within the reactor, or from the radioactive materials contained within the reactors or reprocessing plants themselves. Our need therefore is first to assess the amounts of the various radionuclides that may be emitted into the environment from each stage of the cycle, both during normal operations and in accidents. We then need to link the amount of each radionuclide released (e.g. the becquerels of caesium-137 discharged per year into seawater) with the radiation exposure of people that will result from these discharges (e.g. the man-sieverts collective dose of whole body radiation due to eating fish or crustaceans containing the caesium-137).

Exposure from normal operations

The amounts of the main radionuclides discharged regularly into air or water or in solid form are well established for different types of reactor in different situations, and for typical reprocessing plants. In most cases these amounts are assessed routinely, so that annual or regional variations are known and reliable average values can be quoted.[46, 47]

Similarly the release of radon from mines and of radon or radium from the 'tailings' from uranium mines or mills, and the radionuclides discharged during reactor fuel fabrication have been assessed under typical conditions of operation.

The estimation of population doses resulting from such radionuclide discharges depends on the methods already outlined, of studying quantitatively the ways in which the various radionuclides may enter human food, or be inhaled, or cause external irradiation after being deposited on the ground. The total amount of human exposure for a given discharge will vary considerably, according to such factors as local population densities, wind directions, and the use of rivers for water supply or the sea for fish. More variation is to be expected, therefore, in the local and regional doses, which depend upon these factors, than in the global doses from long-lived radionuclides which are largely independent of them.

A collective dose of about 6 man-sieverts per year, however, gives a representative value for the total exposure of people within a few thousand kilometres of nuclear operations producing 1 gigawatt of electricity.[23] A further 18 man-sieverts would however be delivered world-wide within a period of 500 years, from the global distribution of long-lived radionuclides emitted during that year's output. As summated over this length of time, therefore, a collective effective dose of about 24 man-sieverts will result from each year's output of one gigawatt from nuclear sources (Table 6.2).

Table 6.2
Radiation from nuclear power production

Total collective effective doses (man-Sv per gigawatt year output of electricity)

Source	Delivered: Currently		Delivered: In the far future		
Occupational exposures	25		–		
Aqueous and atmospheric releases					
local and regional doses	6		–		
global doses	4		14		
High level wastes	–		40		
Accidental releases	–		10		
Transport, research, and decommissioning	<1		–		
Total	36	+	64	=	100

Accidental releases

Accidental releases of radionuclides from reactors range upwards from minor discharges of small amounts of material from within the reactor building, without involving any exposure of members of the staff or public, to a possible overheating and 'melt-down' of fuel elements in the reactor core, and escape of highly radioactive fission products and fissile material into the atmosphere.

It is difficult to estimate what might, in the long run, be the average rate of population exposure from a combination of the more common small releases and any rare large ones. This is partly because melt-down has rarely occurred in any nuclear power reactor, and there is little experience of releases involving measurable public exposure on which to base an estimate of what is likely to be a rare event. The release from the Three Mile Island reactor in 1979 involved an estimated total collective dose to all distances of between 60 and 80 man-sieverts, in a year in which the USA output from nuclear sources was about 30 gigawatts of electricity. Even as based only on the year and the country in which this release occurred, and without taking account of other years without significant release, the 'average rate' of exposure would be of 2 to 3 man-Sv per gigawatt year of output.

A substantially larger release occurred at Windscale in 1957, in an early reactor designed and used to produce plutonium for military purposes. The release from this accident has been estimated to involve a collective effective dose of 1250 man-sieverts.[48] The only two known releases from any reactor operation, therefore, have together caused a collective dose of about 1350 man-sieverts in a period (to the end of 1981) during which nuclear power production has totalled some 600 gigawatt years of electricity from civil reactors. As far as simple averages go, therefore, and disregarding any electricity production from military reactors, the average exposure rate on past experience has been of 2.2 man-Sv per gigawatt year's output.

Can we do better by prediction? Two attempts have been made to do so: one in the USA and the other in the Federal Republic of Germany. Both have based their analyses on the fact that any significant overheating of the reactor core is ordinarily prevented by several ostensibly independent mechanical control systems, and will only occur if all these systems fail together. On this basis, tests have been made of the likelihood of mechanical failure, per year of operation, of each system independently. Estimates were therefore derived of the chance of all such systems happening to fail at the same time, or within a time period during which single failures would be detectable.

Both analyses gave assessments on these bases of the probability per reactor year or per unit of electricity output, of releases of different sizes and of the collective doses that would result. Both reached broadly similar results, given the differences in population density assumed in the two cases. The Rasmussen study was based on average population densities at different distances round 66 reactor sites in the USA.[49] It indicated an average collective dose from all, large or small, accidental releases from light water reactors, of 2.5 man-sieverts per year per gigawatt output of electricity (conveniently abbreviated as 2.5 man-Sv per GW(e)y). An uncertainty in this estimate by a factor of three upward or downward was suggested.

The German study,[50] with a ten times greater population density round the 19 reactor sites recorded in the Federal Republic, estimated a collective dose equivalent to 40 man-Sv per GW(e)y. It included a greater number of external causes of accident, such as lightning strike or aircraft crashes, than in the Rasmussen study, and regarded its estimate as a maximizing value.

No annual average rate can properly express the combined impact of very rare large releases and more common small ones. In the long run, however, if many reactors operate over long periods, an average rate would indicate the effect of this source of exposure of the community in causing late effects of radiation, and particularly in contributing to the induction of genetic effects, to which a long-continued form of averaging applies in any case.

The values from both studies may be underestimates to the extent that they assumed that multiple control mechanisms failed independently rather than ever as a result of a single mechanical failure, or human action overriding all controls. On the other hand, both are likely to have overestimated the releases from meltdown in most reactors, in view of the confirmation, obtained during the accident at Three Mile Island, that almost all of the radioactive iodine and other non-gaseous fission products would be deposited within the reactor containment or building, and not discharged to the atmosphere.[51]

Total dose from nuclear power production

Can we estimate the total collective doses that will be received by people now and in the future, as a result of nuclear power production at any given rate?

Part of the assessment is easy, since reliable figures have been obtained in a number of countries on the doses that are actually being received from the various types of occupational exposure that are involved, or that will be received over the next few years or decades from releases from reactors and reprocessing plants.

Part, however, is necessarily more uncertain when it depends on future events such as the long-term frequency of accidental releases, the methods adopted for high-level waste disposal and their efficiency, the adoption of effective ways of controlling releases from mine tailings, the exposures observed while decommissioning large reactors, or the introduction of new methods for removing long-lived radionuclides from effluents.

It is, however, possible to obtain approximate estimates of the likely contributions of these sources of exposure, and some indication of the rate at which people will be exposed as a result of each gigawatt output of electricity derived from nuclear sources. On present evidence it seems likely that this annual collective exposure would reach a rate in the region of 100 man-sieverts for every gigawatt of such electricity (Table 6.2). If, therefore, a country with the present UK average electricity consumption of 0.5 kilowatts per head was deriving half this output from nuclear sources, an annual collective dose of

100 man-Sv would be received by every four million people, implying an average effective dose of 0.025 millisieverts per year or about 1 per cent of that received annually from natural sources. If the consumption of electricity per head of population was at the higher level (of 1.2 kW) current in the USA, and all this supply was derived from nuclear sources, the average annual dose would be 0.12 mSv per year, or 6 per cent of that from all natural sources. (Current rates of consumption of electricity[52] from nuclear and from all sources are given in Table 6.3.)

Table 6.3
Electricity consumption per head of population (1981)[52]

	Kilowatts per person	Nuclear, per cent of electricity generated
North America – OECD (Canada, USA)	1.22	11.8
Western Europe – OECD (23 states)	0.51	16.9
Pacific area, OECD (Australia, Japan, New Zealand)	0.63	12.5
Eastern Europe – centrally planned economies (9 states)	0.50	6.3
Latin America	0.11	0.8
Africa and Middle East	0.06	0.0
Asia (excluding Japan)	0.03	2.9
World	0.22	10.2

The contributions from the various parts of the whole fuel cycle differ somewhat according to the conditions that are assumed to apply, and the bases for their estimation need therefore to be defined in some detail.

1. The value of 25 man-Sv per GW(e)y for all occupational exposures is an average for several countries, the largest contributions to this total coming from reactor operation and reprocessing, with about 10 man-Sv from each. About 1 man-Sv is incurred in the mining and milling of the ore needed to fuel the output of one gigawatt for a year, and an equal collective dose is due to its fabrication into reactor fuel. Research and the development of nuclear techniques have been responsible for about 5 man-Sv per GW(e)y produced, but the figure obtained from different countries varies considerably, and an average value is likely to be rather lower in the future.

2. The collective doses delivered locally and regionally within a few thousand kilometres from nuclear power production plants arise in part from the aqueous and atmospheric effluents, leading to about 4 man-Sv. There is a smaller contribution of less than one man-Sv from mining, milling, and fuel fabrication in the short term, long-term contributions over many generations

requiring efficient control of mine tailings as discussed below (p. 159). The normal contribution from reprocessing is estimated as about one man-Sv. The temporary increase at Windscale (p. 156) due to escape of fission products from fuel rods prior to reprocessing would add a few man-Sv to this figure if it was treated as characteristic of the reprocessing operation, which, however, should not be the case.

3. The global distribution of long-lived radionuclides from a year's output of one gigawatt would cause an estimated collective dose world-wide of about 18 man-Sv by 500 years, largely from the release of carbon-14. The same figure represents the collective dose that would be being delivered globally each year by these long-lived radionuclides if the output of one gigawatt had been derived continuously from present types of reactor depending on nuclear fission for 500 years. The annual dose per gigawatt would have risen to 72 man-Sv if their use had continued for 10 000 years but, in assessing the annual exposure of any world populations from nuclear power production, the value of 18 man-Sv may be considered a maximum contribution from this source on the presumption that power is most unlikely to be derived from fission, rather than from fusion or other sources, even for as long as 500 years.

4. The doses that might be delivered in the far future from any ultimate escape of high level wastes to the biosphere have been assessed by the authoritative INFCE (International Nuclear Fuel Cycle Evaluation) teams reporting in 1979 and 1980. Their estimates varied according to the type of reactor, the use of reprocessing, and the rate assumed for any migration of escaping long-lived radionuclides from repositories to the biosphere. With wastes from reprocessed fuel from conventional light-water reactors, the estimated collective doses[53] ranged from 30 to 50 man-Sv per GW(e)y, and those from fast breeder reactors from 20 to 50 man-Sv per GW(e)y. The value of 40 man-Sv used in Table 6.2 would however be 3 to 5 times higher if fuel were not reprocessed, owing to the much larger amounts of long-lived radionuclides that would be discharged in the wastes, rather than being used as fuel and so broken down to shorter-lived fission products.

5. The estimates that have been quoted above for the average annual collective doses that might be attributable to accidental releases from reactors have been between 2 and 3 man-Sv per GW(e)y from such little experience as there has been to date. The theoretical predictions have been of 2.5 man-Sv from the Rasmussen study and 40 man-Sv from the German one, much of the difference being due to the assumption of a sparse or a denser population distribution in the region of the reactor. For average conditions of population density, and particularly in the light of present evidence on fission product retention within the reactor building following melt-down, a value of 10 man-Sv per GW(e)y appears to be a reasonable, although obviously very tentative, estimate for the average annual dose from all accidental releases.

6. Doses per gigawatt year from other sources are small, with transport of nuclear fuel estimated to contribute a few hundredths of a man-sievert, research and development (excluding their occupational exposures) less than 0.2 man-Sv, and the decommissioning of nuclear plant about 0.3 man-Sv.[25]

7. A year's production of one gigawatt of electricity from nuclear sources is thus likely to cause a collective effective dose in the region of 100 man-sieverts, as the total exposure from all stages of the nuclear fuel cycle.

Taking account of the age distributions of the working and total populations exposed, and of the organs irradiated by the radionuclides released, the collective genetically significant dose is likely to be about one third of this effective dose, for populations of the age distribution normal in most industrialized countries.

OTHER ACTIVITIES OR DEVICES CAUSING RADIATION EXPOSURE

How else are we exposed to radiation in our daily life? Three other sources of exposure need to be reviewed, although the doses received from them are small.

Luminized watches, television sets, and other devices

When watches and clocks used to be luminized with paint containing radium, the penetrating gamma radiation from the radium-226 could cause significant body exposures. Now that the paint can be made equally luminous using artificially produced radionuclides which do not emit penetrating radiation, however, such exposures are trivial. The annual doses to the gonads from radium-luminized watches used typically to be of a few hundredths of a millisievert. The current use of tritium, however, or of an isotope of promethium, has reduced these doses by a factor of about a hundred.

Doses from television sets are also likely to be trivial, although early colour sets could occasionally give significant doses if improperly adjusted. The annual gonad dose from modern sets under normal conditions of use and servicing is, however, unlikely to exceed 0.01 mSv, even to the most obsessive viewer.

Only trivial annual doses are likely to be received also from such devices as smoke detectors, static eliminators, or the baggage scanning equipment used at airports. The x-ray shoe fitting devices, which really could give considerable and uncontrolled exposures to children's feet, have now largely disappeared. Only the use of cathode-ray tubes in scientific work or demonstrations may require care to ensure that substantial radiation is not being emitted from the tube face, as Röntgen found orginally.

Ceramics or glass tinted with uranium or thorium salts may cause limited local exposures, and the concentration of these pigments in spectacle lenses or dentures has in some cases been made subject to control. (No such control was exercized, or indeed needed, when the walls of a children's ward in my own hospital were originally decorated with nursery rhymes depicted in green ceramic tiles. When later converted to a radionuclide therapy ward, Old Mother Hubbard was one day found to be emitting a higher count rate than any contamination that our therapy had ever caused.)

Radiation in flight[23]

As a result of the increase in cosmic radiation exposure with altitude, the rate of about 0.03 *micro*sieverts per hour from this source at ground level increases to 5 microsieverts (0.005 millisieverts) per hour at the altitude of 40 000 ft. usual for long-distance intercontinental subsonic flights; or 12 microsieverts per hour at supersonic altitudes of 60 000 ft.

Because of the greater speed of supersonic flight, however, the dose received in travelling a given distance is ordinarily a little less by supersonic than by subsonic flight. Taking account of altitudes and latitudes of the flight paths, since both affect the radiation rate, it has been estimated that, for example, the London-New York return trip involves a cosmic ray dose of 0.036 millisieverts supersonically, or 0.048 mSv by subsonic flight. Both doses therefore are small, and it would need return trips weekly to make the annual dose from this source equal to that received from natural sources otherwise.

One trip to the moon (and back) would, however, give up to several times the normal annual dose from natural sources. Astronauts who made the first landing received 3.6 millisieverts during their 200 hours away.

Coal and geothermal energy

The burning of coal has two effects on radiation, which work in opposite directions. In the first place, primordial radionuclides are present, in very variable amounts, in different types of coal. They, and their decay products, are released when coal is burnt – radon as a gas, and other radionuclides in the fly ash. Human exposures result from each, and mainly from the deposition of ash round houses where coal is burnt, since the release of fly ash is more fully controlled when it arises in power plants (p. 161).

On the other hand, the carbon present in coal contains less of the radioactive carbon-14 isotope than does the carbon of the air. In the air, carbon-14 is constantly being formed by the action of cosmic radiation (on atmospheric nitrogen, p. 9). In the coal, the carbon-14 that was once present in similar concentration in the vegetation which formed the coal has largely decayed away in the 300 million years since the coal seams were laid down. The effect of burning coal today, therefore, is to dilute the carbon-14 concentration in the carbon which enters our bodies, and so to reduce the radiation dose that we would otherwise receive from this source.

Which wins? It seems likely that the released radioactivity, particularly from domestic uses of coal, increases the population dose by rather more than it is lowered by the reduction of the carbon-14 concentration in air, and that, on balance, the burning of coal may be increasing the world's annual collective dose from natural sources by about 1 per cent.

Energy is also at present obtained on a small scale by 'geothermal' methods. Cold water is pumped down to considerable depths below the earth's surface, at which the rock is at a high temperature. On returning to the surface, the water has not only acquired heat from the rock, but also radon, which is released into the atmosphere. From the mostly small and isolated geothermal plants now operating, the doses to neighbouring communities are small. These local doses could however reach several hundredths of a millisievert annually if large geothermal plants were developed.

SUMMARY OF EXPOSURE FROM ALL SOURCES

This chapter and the previous one have given details of the amounts of radiation received from the various sources to which we are exposed. They have indicated the variation in these amounts that have been recorded in different countries and under different circumstances. As some answer to Chapter 5's opening question, therefore, of how much radiation do we receive, and where do we get it from, Table 6.4 gives the annual collective doses delivered from all radiation sources to people in the UK.[45] Typical exposures from different sources are better illustrated in this way by records from a specified country than by world average values since the latter are very much affected by differences in the use of medical radiology and of nuclear power; and even approximate estimates of exposure from some sources are lacking from many parts of the world.

In the UK, however, detailed estimates have been obtained by the National Radiological Protection Board of current exposures to all significant sources. The results are typical of other countries engaged in similar activities, although the average UK exposure from medical sources is less than half that in some other industrialized countries, and over ten times that likely in many developing countries.

The values are expressed as the collective doses received within the whole population. For sources from which most people receive about equal doses, for example from natural sources, the average individual dose is obtained by dividing the collective dose (man-sieverts) by the size of the UK population (of about 55 million). Thus, the average effective dose from natural sources is 1.86 millisieverts per year, and that from fallout about 0.1 mSv. When, however, the doses are unevenly distributed through the population, as in medical and occupational exposure, the collective doses indicate only the total impact of the source in the population, which is relevant particularly with regard to inherited effects; the individual variations of dose are as described earlier in these chapters.

Table 6.4

Annual collective doses to the UK population

Sources	Effective Man-Sv	Effective (%)	Genetic Man-Sv	Genetic (%)
Natural	104 000	77.8	22 000	88.1
Medical	28 000	20.9	2 600	10.4
Fallout	550	0.41	140	0.56
Occupational				
Nuclear	120		20	
Medical	105		18	
Other	375		58	
	600	0.45	96	0.38
Nuclear Waste				
Aqueous	135		44	
Atmospheric	15		4	
	150	0.11	48	0.19
Miscellaneous	450	0.34	92	0.37
Total	133 750	100	24 976	100

1. Radon in houses contributes about 45 000 man-Sv, on present estimates, to the collective effective dose from natural sources, by deposition of its decay products in the lung. This deposition does not contribute to the genetic dose.

2. The 'other' occupations include various forms of non-nuclear industrial and research activities.

3. The 'annual' doses from nuclear wastes are in fact the collective doses that will be received during the present and all future years, from the present year's aqueous and atmospheric discharges.

Reactors contribute about 10 man-Sv to the effective dose from atmospheric discharges, and reprocessing causes rather less than 5 man-Sv. In each case, about half the dose is delivered within a few years, and half over many generations, mainly from carbon-14.

Reactors cause very little exposure, about 1 man-Sv only, through their aqueous discharges. Of the remaining 134 man-Sv due to aqueous discharges from reprocessing plant, the release of radioactive caesium accounted (in 1978) for 105 man-Sv.

4. If uranium ores were being mined in the UK, a further 3 man-Sv would be added as a result of the occupational exposures incurred in mining and milling the ore sufficient for the electrical output during the year reviewed in Table 6.4, namely about 3 gigawatts from nuclear sources.

5. A further collective dose might be received by the UK population in the future from accidental releases and escapes from high level wastes. On the basis of the estimates given in Table 6.2 and of an output of 3 GW, the former might contribute up to 30 man-Sv (depending on how much of the 10 man-Sv per gigawatt year was incurred within this country). The latter source would add about 2 man-Sv, if the UK population continued to be about 1.5 per cent of the world population. (The collective effective dose world-wide from these two sources, and from the global distribution of components of the liquid and atmospheric discharges, would be estimated as about 200 man-Sv, on the basis of Table 6.2's total of 68 man-Sv per gigawatt year from the three sources together.)

6. The genetic collective doses quoted are the estimated total gonad doses delivered annually to persons (constituting 40 per cent of the population, p. 124) of ages less than the mean age at conception of children.

7

ACTIONS OF RADIATION ON THE CELL, AND THE EFFECTS OF HIGH DOSES

MODES OF ACTION OF RADIATION ON CELLS

It had been evident from the earliest days of radiobiological research that a cell which had been harmed by ionizing radiation might either die without achieving normal cell division, or might survive and divide, but at the risk of transmitting any induced abnormality to its descendant daughter cells.

With increasing knowledge of how cells transmit information to their daughter cells through the chemical structure of their DNA, the ways in which radiation may harm cells have become very much clearer in terms of the damage that may be caused in one or both strands of the double helix of DNA, and the success or failure of the cellular enzyme systems to repair these DNA molecules correctly. Such damage, if unrepaired or inadequately repaired, may prevent the production of viable daughter cells when the cell divides, so that cell death occurs. Alternatively, the damaged cell may be able to divide producing viable daughter cells, but transmitting to them, if so, the uncorrected errors in its DNA, so that any abnormality of function of the damaged parent cell may be manifest in the behaviour of a large family of descendant cells.

It seems likely therefore that the various harmful effects of ionizing radiation on living organisms are largely attributable to one of three basic mechanisms: cell killing; the multiplication of damaged cells derived from the ovaries or testes, resulting in the development of a child with an inherited abnormality; or the multiplication of cells of other body tissues, damaged in such a way that their multiplication is inadequately controlled by normal body processes, so that a cancer results.

Cell death ordinarily becomes significant only if many cells are killed, since most body organs contain many more cells than are needed to maintain the normal function of the organ. Moreover, in some organs cell division will rapidly replace any deficiency. As discussed later in this chapter, therefore, the cell-killing effects of radiation ordinarily become detectable only after high doses, when many cells are killed, except where the function of the organ depends upon the behaviour of a few cells only, as may occur particularly in the developing embryo.

If, however, a damaged cell is able to survive, and divide successfully, the position is quite different. In most cases the effect will be undetectable, since any abnormality of protein synthesis, or deficiency of enzyme production, by a few cells whose DNA programming is faulty will be insignificant as compared to the normal metabolic behaviour of all the other cells of the organ. The two exceptions are of outstanding importance, however, in which the voice of a single cell can be heard. When the affected cell is a germinal cell of the ovary or testis, and is the progenitor of an ovum or sperm which is destined to form a child, all cells of that child may carry the same defect in protein or enzyme formation. The localized chemical alteration in the DNA of a single cell may then be expressed as an inherited abnormality of body metabolism or structure in one or many generations.

In the same way, when a somatic cell of any of the body tissues is changed in such a way that it and its descendants escape from the processes which ordinarily control cell multiplication, the group of cells formed from it may continue to have a selective advantage in growth over the surrounding tissues, and may ultimately increase sufficiently in size to form a detectable cancer, and in some cases to cause death by spreading locally or to other parts of the body.

Both the cell-killing effects of radiation at high dose, and the inherited and carcinogenic effects occurring also at low dose, appear therefore to depend ultimately on the way in which radiation damages the DNA molecules of the cells, and the extent to which such damage is likely to be repaired.

The particles (or electromagnetic waves) of which radiation is composed enter the body tissues with an energy which is characteristic of the radionuclide which emits them or of the type of x rays which are used. As these particles pass through the tissues, they progressively give up this initial energy, mainly by causing ionization of water molecules which lie along the track of the particles. The small unit amounts of energy which are delivered to these ions are, in turn, passed on rapidly to molecular structures in the cell. There is ample evidence that the amounts of energy that are delivered in this way to the DNA molecules of the nuclear chromosomes may be sufficient to break the chain of chemical groups of which each strand of the DNA helix is composed.

There is evidence also to show that, if only a single strand of the double helix of DNA in a chromosome is broken, the cell contains enzyme systems which can correctly repair such a break quite rapidly – typically within a matter of minutes. If both strands of the helix are broken at about the same position, however, correct repair is much less likely to take place. The correct repair of single strand breaks is facilitated by the availability of the second, unbroken, strand to act as a template in the repair process. The broken ends of the first strand are cut out by DNA repair enzymes, and the missing chemical groups are replaced, being identifiable by the groups in the intact chain to which they correspond (see p. 113). In this way the correct sequence of groups is restored, and the information coding the cell's behaviour remains correct.

When however both strands are broken, there is no such template available, and error-free repair is less likely to occur. The strands may either fail to rejoin, or may rejoin with a wrong sequence of the chemical groups which control the normal formation of the cellular proteins. On this view, therefore, most radiation-induced abnormalities of cell behaviour are due to a break in the second DNA strand close to that in the first, and happening before there has been time for the break in the first strand to be repaired.

This double break may happen in two ways. Firstly, if the ionizations are very closely spaced along the track of an ionizing particle, one single particle track may cause 'hits' on both DNA strands. Or secondly, if the distance between ionizations along the track is much larger than the distance between the two strands of the DNA helix, a double strand break at a particular position on the helix will be caused only by single ionizations from each of two separate tracks: hits by bullets from two rifles, rather than by two bullets from the same machine gun.

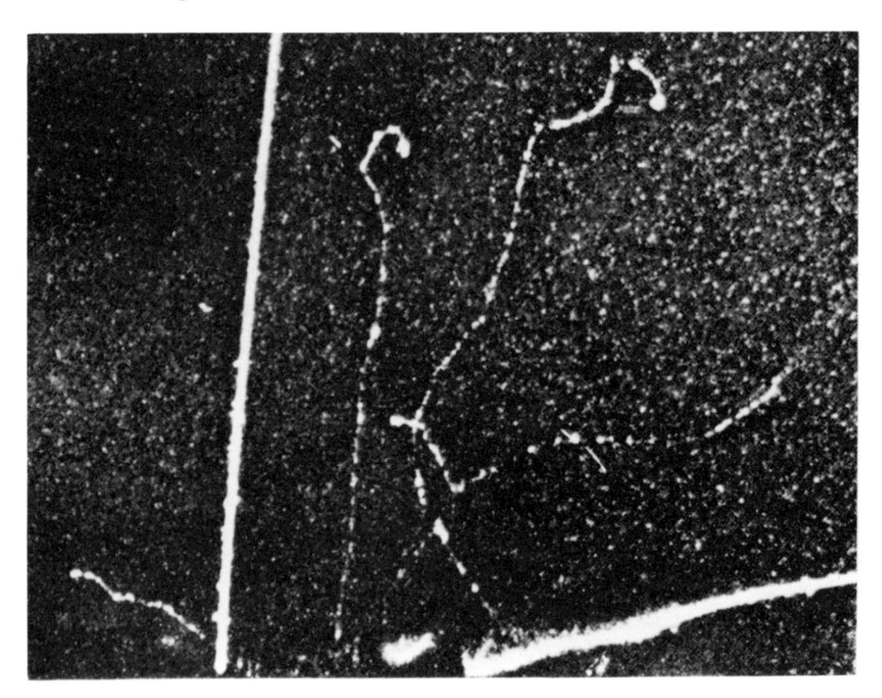

Fig. 7.1. Photographs of short sections of ionizing tracks of alpha particles (of high LET: two densely ionizing tracks; individual ionizations not resolved) and of beta particles (of low LET: four electron tracks, sparsely ionized except at curved terminal sections of the tracks). Reproduced from *Radiations from radioactive substances*[54] by courtesy of the Cambridge University Press.

This interpretation is consistent with the greater biological effectiveness of alpha radiation per unit of energy absorbed, than of x rays or of gamma radiation, particularly at low doses. The heavy alpha particle loses its energy rapidly in its short passage through tissue, and the ionizations that it causes are closely spaced along this short track length (Figs. 7.1 and 7.2). A single track is quite likely, therefore, to cause breaks in both strands. For penetrating x rays or gamma radiation, however, a similar initial energy will only be lost in ionizations distributed along a track which may be ten thousand or more times as long as for the alpha particle; and for most of the track length the distance between ionizations is correspondingly greater, since the average amount of energy lost in each ionization remains about the same. A double strand break must therefore depend on two tracks, each happening to cause an ionization at about the same point on the helix, and doing so within a short time of each other.

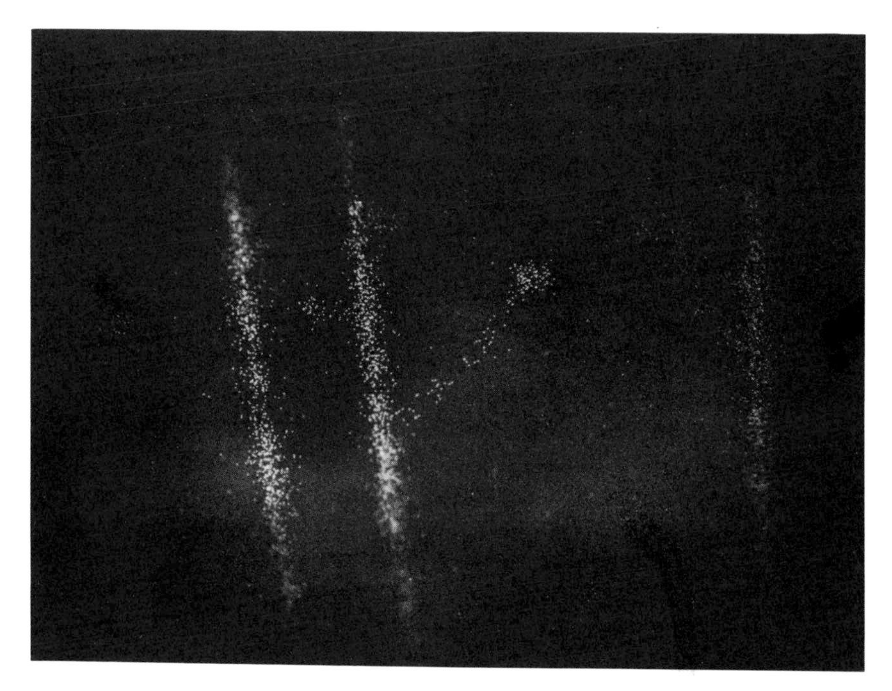

Fig. 7.2. Photographs of short sections of ionizing tracks of alpha particles (of high LET; three densely ionizing tracks; individual ionizations resolved) and of electrons (of low LET, more sparsely ionizing 'delta' tracks, branching from those of alpha particles). Reproduced by courtesy of Dr Michael Marshall, and the Atomic Energy Research Establishment, Harwell.

The densely ionizing tracks of alpha particles are, therefore, likely to cause more double strand breaks per unit of energy delivered than when ionizations only cause such double strand breaks if two tracks coincide closely enough in time and in space, as applies for x rays or gamma rays. Moreover, at a low alpha radiation dose rate, the few tracks emitted still have the same likelihood per track, and therefore per unit dose, of causing a double strand break. At a low gamma radiation dose rate, however, the chance that two tracks, and their ionizations, will happen to coincide in time and space decreases rapidly with dose rate, as the number of tracks per unit time decreases. The ionizations on each track thus become progressively less efficient in inducing double strand breaks as dose rate decreases.

This has important mathematical consequences. For the high LET radiations such as alpha radiation, the number of double strand breaks is likely to be simply proportional to the number of tracks that pass through the tissue; and the amount of biological harm will thus be (linearly) proportional to the size of the dose. With low LET radiations such as x rays and gamma radiation a similar proportionality may apply at very low doses, when the chance of two tracks traversing the same cell at about the same time is very small, and when double strand breaks will occur only occasionally, and then mainly as a result of a close spacing of ionizations in short terminal sections of the tracks (Fig. 7.1). In a higher dose range, however, the frequency of double strand breaks will depend essentially on two separate tracks, each causing an ionization at about the same place and time. In this case the frequency of such breaks, and so the amount of biological harm, should depend on the square of the dose. (The likelihood of any one track causing an ionization at a particular point and time is proportional to the number of tracks passing through the cell per unit time, and hence to the dose. The likelihood of a second track also doing so, at about the same place and time, is similarly proportional to the dose. The chance of the 'double event' is estimated by multiplying together the chances of each single event, and so is proportional to the square of the dose.)

It is in fact found, in radiobiological studies at moderate or low doses, that the probability of cell death or damage caused by high LET radiations tends to be linearly proportional to dose. For low LET radiation, however, this probability more commonly varies in proportion to the square of the dose (a point that we will need to look at in more detail in a later chapter).

The suggestion that biological harm results mainly from uncorrected double strand breaks is therefore broadly consistent with data on the observed effectiveness of high and low LET radiations. It takes account also of much that is now known about the processes of DNA repair, and of the probable size of the target in the cell nucleus within which radiation produces its effects. It is not, however, the only hypothesis that could explain these findings. It has, for example, been suggested that the enzyme systems responsible for correcting breaks or errors in the DNA molecules could cope with such errors when they occur one at a time in a cell, for example as a result of ionizations from separate particle tracks, but not when many errors occur more or less simultaneously, as from a single densely ionizing track. This concept, of local repair facilities being overwhelmed by an increased frequency of damage – a difficulty that any fire brigade knows – could equally well account for the greater effectiveness of densely ionizing radiation than of low LET radiation.

Whatever the exact mechanism of radiation damage to the cell, there is little doubt that the DNA helix itself is the relevant target in producing such effects.

The size of the target has been estimated by using radiations of which the whole track length (of 7 nanometres) is little longer than the distance (of 2 nanometres) between the two strands of the helix. Cells are killed with a frequency that shows that both hits on the target that are required to cause unrepaired damage can result from ionizations within the same short track length. Moreover, the breaking of DNA strands by radiation can be detected and directly measured, using methods which separate chemical compounds according to the size of their molecules. Molecules with the chemical characteristics of DNA, but of sizes corresponding to short DNA fragments, are found soon after radiation exposure. Such fragments become less frequent, however, at longer intervals after exposure during the period in which repair is known to be taking place (as described on p. 117).

It seems likely therefore that cell death ordinarily results if breaks in DNA molecules or chromosomes have remained unrepaired at the time when cell division occurs, so that daughter cells with the necessary content of chromosomal material cannot be produced; or, alternatively, that inherited defects or cancer may result when the uncorrected errors in coding are compatible with continued cell life and division.

CELL KILLING

A great deal is known about the effects of radiation in causing the death of cells of different plant or animal species, or of such cells grown in culture. Since these deaths result from a failure of the normal reproductive capacity of the cell and since cells of different tissues ordinarily divide at different rates, the frequency with which cell deaths occur after radiation will depend in part on the type of cell irradiated and its normal rate of division.

The probability that a cell will be killed by a given radiation exposure depends also upon the stage in the cell's normal cycle of growth and division at which the exposure occurs. Thus it is found that cell death is much more likely to result from irradiation delivered while the cell is dividing than during the more resting phases of the cell cycle. It follows also from this that a given radiation dose will kill a greater proportion of the cells of a tissue in which cells are dividing frequently, than in a tissue in which cell division is rare. This, of course, is one of the reasons why many cancers can be controlled or successfully destroyed by radiotherapy, without undue damage to the surrounding normal tissues. In the more rapidly growing cancer, most or all of the cells can be caught in the radiosensitive stage of cell division by the successive exposures used in the course of therapy. In most normal tissues that are exposed during treatment, cell division is infrequent and cell killing is less common.

This sensitivity of dividing cells has, however, an obvious and important bearing on the effects of high doses delivered to the body over short times.

If a dose of a few sieverts is delivered in this way, most cells may be killed in those tissues, such as the bone marrow or the intestine, in which the cells divide frequently. The necessary supply of new daughter cells to the blood from the bone marrow, or to the lining of the intestine, may therefore be cut off, with severe or fatal results.

If however the same dose is delivered over a longer period of time – for example of weeks rather than of hours or minutes – the very fact that cell division is frequent in these sensitive tissues will prevent any detectable harm from resulting. The amount of cell killing is offset by the high natural rate of cell replacement, and no significant shortage of cells occurs in blood or gut.

In other body tissues, in which cell division is infrequent, the opposite applies. Exposure to a few sieverts ordinarily kills too few cells in these more radio-resistant tissues to cause any detectable harm, whether the dose is delivered over a short or a long time. The low natural rate of cell replacement in these tissues, however, means that, after higher doses delivered over long times, it is these tissues which are liable to show the greater damage, since killed cells are not adequately replaced. The deficiency of cells accumulates, and may finally lead to significant impairment, or failure, of the function of these tissues or organs, whereas in the more radiosensitive tissues, cell replacement has kept pace even with the greater rates of cell loss.

We must therefore review two categories of radiation injury which are associated with cell killing. In the first, an early or so called 'acute' damage results from the exposure of radiosensitive tissues to a few sieverts in a short time. If the whole body is so exposed, the effects may prove fatal within a few weeks or even days. In the second, much larger doses – typically of 20 sieverts or more – delivered to other body tissues can cause failure of the tissues irradiated in this way. These effects ordinarily develop as late effects of exposure and may be seen in tissues which have needed to be heavily exposed during radiotherapy given some years previously.

In both cases, of the acute effects and of these late effects of radiation, it is recognized that there are 'threshold' levels of dose below which no harm occurs, but above which the harm becomes both more frequent and more severe. This follows from a dependence of much if not most of the harm upon cell killing. For doses which kill few cells compared with the total number of cells in an organ, no harm is detectable, since most organs contain a large reserve capacity of cells, and the normal functions of the organ can be maintained by the cells which are unaffected by the exposure. After all, half of the body's total kidney or lung tissue can when necessary be removed surgically without appreciable detriment if remaining parts of these organs are normal. If however, a sufficient number of cells are killed, the function or structure of the organ will be impaired; and the greater the cell killing the greater the impairment is likely to be.

Much of the serious damage that is caused by high doses of radiation is understandable in this way in terms of the number and types of cells which are killed at these dose levels. Various secondary effects of cell destruction also contribute, however. For example, the swelling, ulceration, and inflammatory reactions that can follow any abrupt tissue injury may occur following large exposures at high dose rates. Or, if considerable numbers of cells are killed as a result of prolonged exposure at rather lower dose rates, their place may be taken by fibrous tissue, with consequent scarring and impairment of the structure as well as the function of the organ. Changes in the blood-vessels and the blood supply of chronically inflamed or damaged tissues are also common, whatever the cause of the injury. The importance of any particular amount of cell killing in an organ will depend very much on the type of response that occurs in that organ to such damage, as well as on the function of the organ in maintaining normal health and activity.

These results of high radiation exposures, which are likely to occur whenever a threshold dose has been exceeded, are commonly, if awkwardly, referred to as 'non-stochastic' effects. This is intended to contrast with the hereditary or carcinogenic effects of radiation, of which the dose determines a probability rather than a certainty of the effect occurring, and which are termed 'stochastic'.

LETHAL EFFECTS OF RADIATION

In Hiroshima and Nagasaki, many thousands of people will have been killed by radiation, but many tens of thousands more by the combined effects of the radiation, burns, and blast pressure created by the explosions. For this and many other reasons, no clear evidence was obtainable either of the amount of radiation necessary to cause death, or even in any detail of the ways by which radiation exposure alone can cause death.

Since then a total of some 17 such immediate or early deaths are known to have occurred, as a result of two kinds of accident. Eight of these deaths have been due to accidental high exposures received in the course of research into the physical processes involved in nuclear reactors,[23] and particularly when fissile materials came together at distances, and in quantities, which exceeded the critical amounts required to start a small-scale fission reaction.

The other nine deaths occurred following occasions when capsules of radioactive material, which had been used for industrial radiological purposes, had been left inadequately secured and shielded. They were found, usually by children, who took them back to their homes without anyone in the family recognizing their nature and their menace.

In one of these episodes, in Mexico in 1962, a 5 curie source of cobalt-60 was being used in the radiography of metal structures to detect cracks or defects.[55] It was found on a building site by a child of 10, taken home in his pocket, and kept in a wooden cupboard in his house. He, and three of the four other members of his family, died during the ensuing months from the high doses they had received from this large and virtually unshielded source.

In a second episode, in China during 1963, a 10 curie cobalt-60 source which was being used for irradiating seeds, was similarly found by a child and taken to his home.[56] The boy and his brother, who were estimated to have received high average whole body doses of 40 and 80 sieverts, died despite active treatment. Three other members of the family, who are likely to have received doses of between 4 and 8 sieverts, survived, as did an uncle who visited the house only for one night.

The third of these tragedies happened in Algeria during 1978, when a 25 curie industrial radiographic source of iridium-192 caused the death of one of the 22 people who had been similarly exposed at home. In this instance, as in the others, the nature and cause of the developing illnesses were recognized by local medical staffs, despite the highly unfamiliar and unexpected character of the symptoms; and the active treatments given will certainly have saved the lives of some of those who were at risk.

These events, however, and those that resulted from the criticality accidents in nuclear research, have at least given vital evidence on the way in which high doses of whole body radiation cause illness or death in man, and approximately the dose levels at which they may do so. In this way they have indicated the lines which treatment should follow, and have amplified what was already known from the effects of partial body exposure in radiotherapy, and from studies of the effects of radiation in animals.

Some of the injuries observed were due to very high doses – probably of many tens of sieverts – received locally by parts of the body with which the sources had been in close contact – for example in the Mexican boy who had carried the capsule home in his pocket. Although extensive necrosis of the skin and underlying tissues was caused in a number of individuals in this way, the local damage could usually be adequately treated by surgical means and did not in itself cause death.

In certain of the fatalities from the criticality accidents, when very high doses of some tens of sieverts are likely to have been received in a short time, death occurred within a day or two of the exposure, with symptoms indicative of overwhelming damage to the brain and the nervous system, or failure of the heart.

After less extreme doses, in the region of 10 to 30 sieverts to the body as a whole, death has occurred at between 1 and 2 weeks after exposure, with vomiting and diarrhoea, intestinal bleeding, and shock, indicating severe gastro-intestinal damage. A victim of one of the criticality accidents, and the two brothers who died following the Chinese accident, died in this way at between 9 and 12 days after the initial exposure.

Otherwise, however, when brief exposures of less than 10 sieverts have proved fatal, death has occurred at a month or two after the exposure, with relatively few symptoms during the first week of this period. From the time of the exposure, however, the number of circulating cells in the blood begins to fall – not from any continuing action of radiation on these cells themselves, but as a result of destruction, at the time of the exposure, of most or all of the 'stem' cells in the bone marrow, of which the repeated division normally maintains a steady supply of the corresponding daughter cells to the blood. The effects of the exposure are at this stage therefore a consequence of the greater likelihood that radiation will kill cells which are dividing than those which are not.

The important stem cells in the bone marrow are those which, by their division, produce three essential constituents of the blood: the white cells, which are involved in combating infections; the blood platelets which prevent haemorrhage by plugging leaks in blood vessels; and the oxygen-carrying red cells. None of these products are rapidly dividing cells, so none of them are radiosensitive in the way that the stem cells are. Indeed, the red cells have no nuclei which would enable them to divide, and the platelets are not cells anyhow. Radiation therefore reduces the supply, and not the existing number, of these three essential blood constituents. In consequence trouble develops, not when the supply of new elements is cut off at the time of the exposure, but when the current number of each of these elements – red, white, or platelets – is sufficiently reduced by the normal processes of wastage or loss of them from the blood.

For example, a red blood corpuscle normally survives for about 120 days before it is dissolved or removed from the bloodstream. In the absence of replacements, therefore, the red cell count will fall by about 1 per cent of its initial value per day, although any abnormal causes of blood loss will accelerate the consequent development of anaemia. Similarly the various types of white cells, if not replaced from the marrow or the lymph nodes, decrease in number at rates corresponding to the normal survival times of the different types of these cells; one type, the lymphocyte, falling to low levels within a few days. As regards their function in combating infections, however, critically low levels are likely to be reached by 3 or 4 weeks from exposure, after which any recovery that is possible, by renewed supply from the stem cells, is likely to show. The fall in blood platelets becomes important also at about a month.

This sequence of events may be affected by the consequences of cell killing at other sites of rapid cell division than in the bone marrow. In particular, ulceration may develop in the intestine, either from depletion of cells forming the intestinal lining, or more probably at the site of the clusters of dividing white cells which are always present in the gut wall; and such ulcers give risks of blood loss or infection. Similar risks will result from any skin ulceration that develops, although such ulceration is unlikely unless the exposure had been much greater at some positions than to the body as a whole, or had resulted from weakly penetrating radiation which affected the skin more than the other body tissues.

The typical course of radiation sickness leading to death from bone marrow failure therefore is likely to be of weeks in duration rather than of days or hours; initial and transient gastro-intestinal symptoms of nausea and vomiting would not be life threatening in themselves. This can allow time, given adequate medical facilities, to establish treatment and possibly reversal of the potentially lethal factors. Of these, the two most important are infections and blood loss. Blood loss, whether from the gut, from heavily irradiated skin, or resulting from platelet deficiency, would require compatible blood transfusions which might need to be large and continuing if marrow recovery was long delayed.

The prevention of infections is more difficult, since the decreased immunity that results from reduced white cell production can allow infections which are ordinarily trivial to develop and spread dangerously. Two special methods are likely to be important, in addition to the use of any appropriate antibiotic and the treatment of locally damaged areas of skin. In the first place, an attempt can be made to prevent even the commonest and ordinarily mildest sources of infection from reaching the patient, who must therefore be fully enclosed within a sterile compartment, into which ideally only sterile food, dressings, equipment, or gloved hands should penetrate.

The second method is by bone marrow transfusion, in the hope that the marrow cells of a compatible donor may colonize the depopulated bone marrow of the patient, and provide a sufficient, although alien, supply of white cells and perhaps other blood constituents.[57] In the ordinary way, such an attempted graft of marrow cells would probably be rejected by the immune system of the recipient. If this immune reaction has been sufficiently depressed by the radiation exposure, however, the chance of a successful 'take' is increased, and bone marrow grafts have sometimes been effective in cases of severe radiation exposure, the marrow cells injected intravenously into the recipient's bloodstream becoming established and proliferating in his bone marrow.

The size of the lethal dose of radiation

From what has been described, it is evident that no exact figure can be given for the amount of radiation that will cause death. No exact estimate could in any case be expected, since the necessary exposure will vary somewhat, although probably not greatly, in different individuals. The amount will also depend on the medical treatment that can be given. It will certainly depend upon whether the whole body is uniformly irradiated, and on the duration of the exposure, since a given dose will be considerably less dangerous if delivered during a month than during a minute. A brief uniform exposure of the whole body of between 3 and 4 sieverts is, however, likely to cause death within a month in about half the people so exposed, and in the absence of the more highly specialized forms of medical treatment. After this period of about a month, further deaths from the cell killing effects of a brief exposure are less likely, although late effects from cancers developing within the following 30 or 40 years could be expected to add to these early deaths of 50 per cent of those exposed at this dose level, late deaths from cancer in a further 2 or 3 per cent.

The estimate of between 3 and 4 sieverts as the median acute lethal dose, coldly termed the LD50-30 (the Lethal Dose in 50 per cent of people within 30 days), is based partly upon the frequency with which death has resulted from the various accidental exposures that have been described, and in which some approximate estimate could be made of the whole body exposures received. Its value depends also, however, on knowledge of the rather narrow range of dose, in various mammalian species, between that below which no deaths occur, and that above which all exposed animals will die. The exposure for 50 per cent lethality in man is likely therefore to be only moderately higher than that at which the risk of death is 10 or 20 per cent, and the latter dose can be estimated, at least approximately, from the accidental exposures that have occurred. The way in which the early mortality increases with dose in a number of other species suggests that few such human deaths will occur following brief whole body exposures at less than 2 sieverts, but that there will be few survivals, in the absence of highly specialized medical treatment, after doses of over 8 sieverts.

These necessarily very approximate estimates apply to the risks of brief exposures. In the case of prolonged exposures extending over weeks rather than hours or minutes, the size of dose required to produce any particular amount of damage or risk of death is greater, perhaps by a factor of 2 or 3, owing presumably, at least in part, to the proliferative capacity of the stem cells surviving at each stage of the exposure.

EFFECTS OF CELL KILLING ON ORGAN FUNCTION[23]

Doses of a few sieverts are unlikely to cause significant impairment in the functioning of most organs in which cell division is infrequent, since the function of the small numbers of cells that are killed by such doses, whether delivered in short or long periods of time, will ordinarily be maintained by the large numbers of cells in the organ which survive. Effects are only likely to be detectable after doses of over 10 sieverts delivered in short periods, or of over 20 or 30 sieverts in longer ones.[58]

Following whole body exposures of brief duration therefore, effects on these organs are not prominent, since the damage to the bone marrow and other more radiosensitive tissues will dominate events and determine the outcome. Early or acute effects of brief exposures of most organs are likely to be seen only when these individual organs are selectively and heavily irradiated. This has occurred following radiotherapy, as for example when fibrosis of the lungs has developed within one or two years of treatment involving the lung fields at high dose. It may also follow the intake of high activities of radionuclides which become concentrated selectively in particular organs. This effect has been used therapeutically in the administration of large amounts of radioactive iodine in order to achieve the suppression of the normal thyroid gland function which is necessary as the first step in treatment of disseminated thyroid cancers, or in reducing towards normal the activity of an overactive gland. In the latter case, the gland's activity is usually reduced to a normal level by amounts of radioiodine delivering doses to it of 30 to 40 sieverts. In the former, the necessary complete and immediate destruction of the gland is ensured by doses of several hundred sieverts.

Apart from such high local doses that may sometimes be received in the course of therapy, or following accidental exposure of parts of the body in close proximity to unshielded radioactive sources, significant damage to the function or structure of most body organs is likely only to be seen as a late result of the prolonged intake of substantial amounts of any radionuclides which are selectively concentrated in these organs, and if doses of some tens of sieverts are eventually exceeded.

Four exceptions should be noted, however, in which higher sensitivites are to be expected.

The embryo

In most tissues, the loss of a small number of cells is completely without detectable effect. In the developing embryo however, at a stage in development at which a few cells only may be responsible for the subsequent formation of whole organs or parts of the body, even the limited cell killing or damage which may result from small doses, for example of the order of a few hundredths of a sievert, may involve a significant risk of causing serious malformation.

At later stages of embryonic development, when the primordial organs already consist of a substantial number of cells, the destruction of limited numbers of such cells might still have severe consequences if the organ were of the complexity and differentiation of the brain or nervous system. Indeed, serious mental retardation has occurred in some Japanese children who were exposed to radiation before birth from the bombs in Hiroshima and Nagasaki at a stage in their development at which the brain will have been forming.

It is not likely that the normal formation of any body organ or structure ever depends solely on the survival of a single cell. Even at early stages of embryonic growth, other cells have the potentiality to take on the developmental role of any single cells that are lost. It is probable, therefore, that a threshold dose would always need to be exceeded before all cells from which an organ could develop were killed. Since, however, damage to relatively few cells may cause malformations, the threshold dose for production of such effects may be low. In this sense the risk to embryonic tissues is more akin to that of causing inherited abnormalities by irradiation of germinal tissues than it is to the risks from cell killing in general, which become significant only when a large threshold dose is exceeded.

The testis

In the testis, sperm cells are continuously being developed from the division of stem cells through a series of stages until the fully developed spermatozoa are formed and stored within the organ. In most of these developmental stages, including the original stem cells, the developing sperm cells are not of high radiosensitivity; and the adult sperm, which by now are non-dividing cells, are highly radioresitant. At one of the earlier stages in their development, however, the cells, then called spermatogonia, are more radiosensitive and quite low doses of less than one sievert may cause the sperm count of live spermatozoa in the ejaculate to be decreased. This fall in numbers would occur at some weeks after such an exposure, as the decrease in spermatogonia causes a shortage in supply of adult sperm.

Provided that enough of the spermatogonia, or of other more radioresistant stem cells, have survived however, the sperm count returns to normal as spermatogonia and cells of later stages are replaced by stem cell division. This restoration, however, and the return of full fertility might in some cases take over a year to be reached. After higher doses of several sieverts, insuffient stem cells might survive to repopulate the semen-forming tubules of the testis, in which case permanent sterility would result.

The hormone-forming cells of the testis are unlikely to be affected at such doses, so that potency would not be reduced.

The ovary

The ovary differs from the testis in that, whereas the testis contains what is in effect a continuous production line of developing sperm, the ovary acts only as a storehouse of the cells – the oocytes – from which the ova are formed. Since the adult ovary contains no stem cells, therefore, the initial supply of oocytes which is present at birth provides all the ova that will be released during the successive ovulations of reproductive life, so that the number of remaining oocytes decreases progressively with age. Individual oocytes, however, mature at different times during adult life, and it is the mature ones that are released as ova during ovulation.

Mature oocytes are rather more radiosensitive than developing ones. Doses of a few sieverts may therefore kill mature oocytes, and so impair fertility temporarily, without killing immature oocytes and impairing it permanently. Permanent sterility would be likely at higher doses, however, if oocytes at earlier stages of maturation also were greatly reduced in numbers.

Because of the normal reduction in number of remaining oocytes in later life, both temporary and permanent reductions in fertility are likely to be caused at rather lower doses in the middle-aged than in the young.

The lens of the eye

A fourth instance of significant late non-stochastic effects being induced at doses below 10 sieverts concerns the lens of the eye, with the possibility of causing lens opacity or cataract. The formation of cataract is here due to the fact that, if cells in the tissues round the edge of the lens are killed, the dead cells become displaced to a central position at the back of the lens. If sufficient numbers of such dead cells accumulate in this position, on the central line of sight of the eye, they may form a large enough area of opaque material to impair vision, and so constitute a cataract. Smaller numbers cause no detectable impairment of vision, but may be found to be present by special methods of examination, for example by inspection of the eye by the slit lamp.

There is evidence from radiotherapeutic experience, for example in treating tumours of the eye or eyelid, that exposures of the lens at doses substantially above 10 sieverts can be followed by cataract – that is, by opacities which cause detectable impairment of vision. There is some indication also, however, that lower doses, in the region of 6 or 8 sieverts, can cause smaller opacities which do not impair vision but which may progress, without further radiation exposure, to a size at which vision is impaired.

Other body tissues

We are therefore seeing a threshold for severe and lasting effects in the function of the testis, ovary, and lens, at doses which, if delivered to the whole body uniformly in a short time would commonly cause death; and effects on the embryo from much lower doses, if delivered at certain critical stages in its development.

For other body tissues the thresholds for any clinically detectable effects appear to be substantially higher, with values over 10 Sv for doses delivered in a short time, and probably over 20 or 30 Sv for doses delivered over periods of months or years. The evidence for the latter is derived largely from experience in radiotherapy, in which treatment is ordinarily continued over a month or more, when some normal tissues are likely to be included in the treatment field, or irradiated at high dose because of a possible spread of cancer cells into them. Many patients who have been so treated for malignant disease have been followed closely for long periods of time, so that the late effects of known exposures on normal tissues are well identified, and appropriate limits of irradiation established.[58]

One qualification needs to be emphasized, however, with regard to this clinical experience. The threshold doses quoted are those below which no substantial impairment of organ function is thought to occur; or, more probably in some instances, no effect which is regarded as an unacceptable complication in the treatment of a life-threatening disease. It is evident, however, that for any effect which depends for its severity on the numbers of cells that are killed by radiation and are not replaced by subsequent cell division, there are likely to be different degrees of impairment at doses both above and below the threshold at which normal function is maintained under ordinary conditions of life. Not much is at present known, however, about how well different organs cope with different levels of challenge after only small proportions of their normal cell populations have been killed, whether the increased challenge takes the form of increased amounts of chemical substances to be metabolized or excreted, or of different amounts of exertion or other forms of stress.

In this connection it is reassuring that, even after the loss of half of their cell mass, many organs appear to maintain an essentially normal function under most conditions, and even under conditions of substantial stress, as judged by the general experience and the results of investigations following, for example, the removal by surgery of one kidney, one lung, one lobe of the thyroid, or a substantial part of the liver.

It is reassuring also that, in the long and careful studies that have been made of various populations exposed to radiation from medical treatment as well as in Hiroshima and Nagasaki, no increases have been found in the incidence of non-malignant diseases of the kinds which would be expected if doses of up to several sieverts caused significant impairment of the organs so exposed.[22]

One observation, however, illustrates the need for some caution about this question. It has been mentioned that the activity of the thyroid gland is rapidly and completely abolished by doses of a few hundreds of sieverts. It is known also that the gland's function can be moderately reduced, to relieve the severity of symptoms in some forms of pulmonary or cardiac disease, by doses of 20 to 30 sieverts, substantially lower doses being ineffective in this respect. It has however been reported that several individuals, estimated to have received thyroid doses in the region of 3 sieverts, from fallout in the Marshall Islands many years previously, had a normal thyroid function and essentially normal blood levels of thyroid hormones, but had somewhat raised blood levels of the pituitary hormone responsible for maintaining thyroid activity.[59]

The situation would seem to be one in which, after moderate doses and with limited amounts of cell killing, adjustments of body control mechanisms may be fully sufficient to prevent any consequent impairment of organ function or body activity, perhaps in much the same way as acclimatization to life at high altitudes involves blood changes which compensate for the reduced oxygen pressure in the air. It must however be recognized that thresholds for these non-stochastic effects are necessarily set in terms of particular, and to that extent arbitrary, levels of detectable impairment, and that these levels need therefore to be reviewed in relation to the amount of detriment that they may cause under any normal or adverse conditions of life.

We will need to return to the prevention of harmful effects that may result from cell killing when we consider the criteria required in efficient radiation protection. Meanwhile it appears that, except in the important case of the embryo, doses of the size needed to cause detectable effects from cell killing – doses of several sieverts or several tens of sieverts – are most unlikely to be incurred except sometimes in the course of radiotherapy for serious disease, or from accidental exposure to unshielded sources of high activity.

In occupational exposure, annual doses to the whole body typically average about 0.005 Sv per year or less, and rarely average more than 0.03 Sv per year; doses of 2 Sv or more are rather unlikely to be received during the working life of any individual, either in the whole body or in most individual organs. In members of the public, the annual dose from natural sources normally average about 0.002 Sv per year, to which medical and all other man-made exposures may add another 0.001 Sv per year – again as an average. A lifetime total dose of over 0.5 Sv to the whole body is unlikely in any individual, although the lungs may receive 1 Sv or more where radon levels are high in houses, and other tissues may be exposed to several tens of sieverts in the course of any radiotherapy that the individual receives.

In general, therefore, at the highest lifetime doses likely to be received either occupationally or amongst the public, it must be the carcinogenic and the hereditary effects of radiation, rather than those attributable to cell killing, that can cause harm and that require particular evaluation.

8

GENETIC EFFECTS OF RADIATION

TERMINOLOGY

In describing the inherited and the carcinogenic effects of radiation, we have an initial problem of terminology. Both effects are ultimately attributable to damage to the DNA structure of the chromosomes, and so to an altered behaviour of the genes of the affected cells. In this sense, both are forms of 'genetic' harm, whether in germ cells causing the inherited effects or in cells of other tissues causing cancer. The term 'genetic' is, however, now so widely applied to inheritable defects, and indeed to their study by geneticists in the science of genetics, that it will commonly be used in this way in the following chapters.

THE QUESTION OF THRESHOLDS

In reviewing the non-stochastic or mainly cell killing effects of radiation in the last chapter, we needed to assess the differing severities of these effects, and the threshold doses above which they were liable to be caused. In this and the next chapter the requirement is different. Here we will need to assess the frequency with which hereditary abnormalities or cancers are induced per unit dose, or at any particular dose level. If such an effect is induced, however, its severity does not depend on the size of the dose which caused it.

Moreover, it cannot be assumed that there is any threshold dose below which these effects will never be caused. It is certainly true that the enzyme systems present in the cell nucleus can and do repair much of the damage to the chromosomal DNA which is caused by radiation. It is also arguable, as already mentioned, that the efficiency of repair may well depend upon the dose or dose rate received by the cell. It is tempting to suppose that these repair mechanisms might completely repair damage that was being caused slowly at low dose rates, but that their capacity might become saturated at a higher rate, so that only the damage inflicted at these higher dose rates would be incorrectly repaired.

This argument would be persuasive if exposure at low dose rates caused few ionizing tracks to pass through the nucleus per hour, while rather higher dose rates caused many. In general, however, even at a moderate dose rate of one millisievert per hour, few such tracks will pass through any one cell nucleus in each day. The difference between two low dose rates depends much more therefore upon the number of cells through which any track passes, than on the number of such traversals per cell. It cannot be assumed that the capacity of the enzyme systems in one cell would be significantly affected by whether many or few distant cells were involved in repair at the same time. While therefore the efficiency of DNA repair may be dependent on dose rate at higher doses, this would not constitute an argument for the existence of a threshold at dose rates at which two tracks were unlikely to pass through any one cell nucleus in the period, probably of a few hours only, during which any possible DNA repair was likely to be completed.

In assessing the amount of harm that may be caused by many man-made sources, we are not in any case considering whether there may be a threshold of safety at extremely low dose rates. Natural sources of radiation are already causing a dose rate of several microsieverts per day. The question therefore is whether there might be a threshold at some dose rate a little higher than this, which would protect against artificial sources which only slightly increased the rate of exposure from natural sources; and there is no evidence that this is so. Nor is there any reason to imagine that the chemical components of the cell could somehow distinguish between ionizing tracks resulting from natural and from artificial sources, accepting the former as harmless goodies and only the latter as baddies which could cause damage.

Evidence has been produced, however, to suggest that there might in fact be a threshold dose below which certain types of cancer failed to develop, even though there was no threshold for the cellular changes which could initiate such a cancer. There is ordinarily a long latent period of many years between the exposure which causes the relevant cellular transformation, and the appearance of a cancer resulting from the proliferation of the transformed cell. For bone cancer there is some evidence to indicate that this latent period is longer following small doses than after larger ones.[60] If, below a certain dose, the minimum latent period exceeded the natural life expectancy of the person who received the dose, there could be a 'practical threshold' at the age at which he received the dose, below which a cancer which had been initiated by the relevant cellular transformations would fail to develop during his lifetime and so would never cause symptoms or harm.

The practical implications of this possibility might be far reaching in conditions of very low dose exposure, but are still by no means clear even for bone cancers. In general, therefore, it seems necessary to assume that, for most if not all cancers and for all inherited abnormalities, harmful effects may follow even the smallest doses, although the frequency of effects due to small increments above the doses received from natural radiation sources are likely to be very small.

INHERITED ABNORMALITIES

The detection and the early studies of the genetic effects of radiation gave no indication of any significant threshold for these effects. Hermann Muller found, in 1927, that abnormalities were induced by x rays in the progeny of male fruit flies that had been irradiated at rather high doses of 34 Sv.

Further experimental work on genetic effects in fruit flies at progressively lower doses tended to confirm the supposition that the number of mutations induced might be proportional to dose down to low doses, without threshold. The frequency of mutations expressed in the progeny was shown to remain linearly proportional to the exposure down to a dose of 3 Sv in 1930, 0.25 Sv in 1949, and 0.05 Sv in 1961. These doses are still high in relation to an annual dose of 0.001 Sv from natural sources, but gave no grounds for assuming a threshold. A similar linear proportionality was found in 1951 down to doses of 0.4 Sv in the mouse – the rather higher doses used reflecting the greater problems in studying tens of thousand mice than tens of thousand fruit flies. Although it was later found that a given dose was less effective, by a factor of about three, in inducing mutations in mice if delivered at a low rate than at a higher one, the frequency of effects per unit dose remained about constant at each of the dose rates. This relationship between frequency of mutations and dose is even more easily examined at low dose in plants than in fruit flies; and in Tradescantia, the spiderwort, mutations have been shown to be produced by doses of only about 3 millisieverts.

Before describing the genetic risks of radiation, and the methods of estimating their magnitude, it may be useful to outline the normal chemical mechanisms of inheritance, and the extent to which damage to these mechanisms can be repaired by cellular processes.

In human tissues, there are normally 23 pairs of chromosomes in the nucleus of every cell, one member of each of these pairs having been derived from the father and one from the mother (Fig. 8.1).

When a cell divides, each of the chromosomes in it duplicates and divides also, so that each of the daughter cells contains an identical copy of the same 23 pairs of chromosomes. This applies equally to the dividing cells of body organs such as the kidney, and to those forming the germ cells of the ovary and testis. In the case of the germ cells, however, the process is different at one stage in the final development of the ovum and the sperm. At this stage, each daughter cell receives a copy of one member only of each pair of chromosomes, instead of a copy of both. This reduction in chromosome numbers, from 23 pairs in the dividing cell to 23 only in each daughter cell, is obviously needed at some point in the process of sperm and ovum formation, since otherwise the fertilized ovum would find itself equipped with 46 pairs.

The cells of the developing embryo thus receive half only of the genetic information that determined the cellular metabolism of the father, plus half of that from the mother. This incidentally ensures that members of a family differ, not only from their parents, but also – except in the case of twins developed from a single fertilized ovum – from each other. It also gives the basis for the inheritance of the dominant or recessive characteristics referred to later.

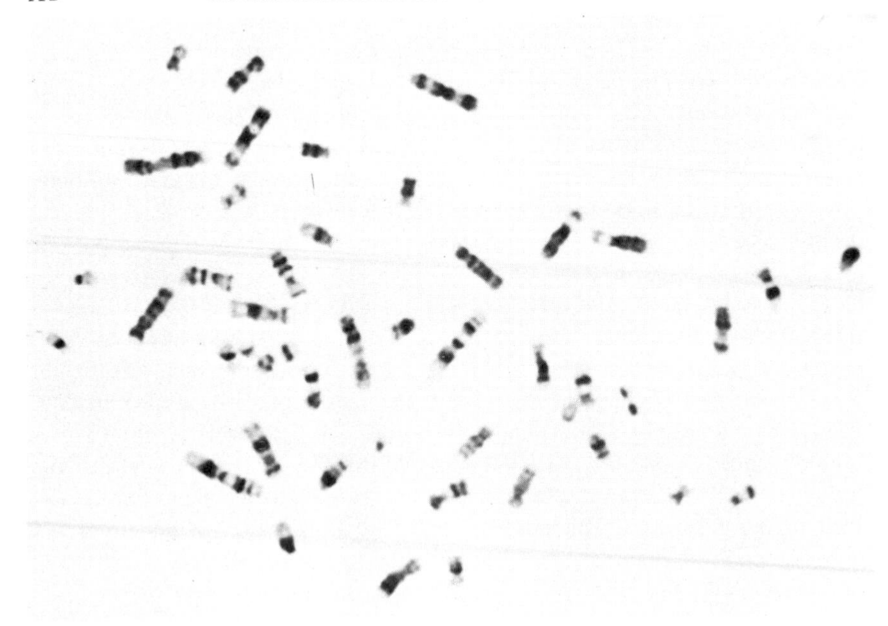

Fig. 8.1. Human chromosomes (normal male, metaphase) from blood culture, G banded (by trypsin technique GTG). Reproduced by courtesy of Mr George Brecken and of the MRC Radiobiology Unit, Harwell.

Each individual chromosome contains two continuous threads, or strands, of DNA, which are ordinarily wound round each other in the form of the familiar double helix. These strands separate during cell division, so that each can become

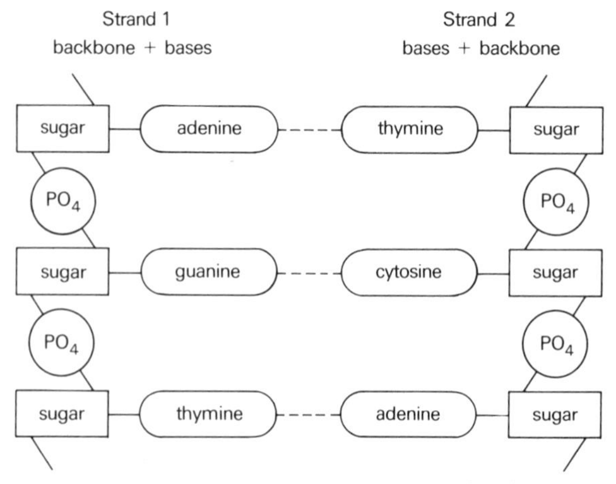

Fig. 8.2. Diagram of a section of both DNA strands of a chromosome and of the cross linkage between them through their bases.

duplicated ready for transmission to the two daughter cells. In this duplication each old strand acts as a template for the chemical synthesis of the necessary new strand.

The chemical structure of the DNA strands makes this possible. Each strand has a continuous backbone of alternate sugar and phosphate groups, chemically linked in series (Fig. 8.2). To each sugar is attached one of four bases: adenine, thymine, guanine, or cytosine. The two strands are linked to each other through the successive pairs of bases, adenine on one strand always occurring opposite thymine on the other, and guanine always opposite cytosine.

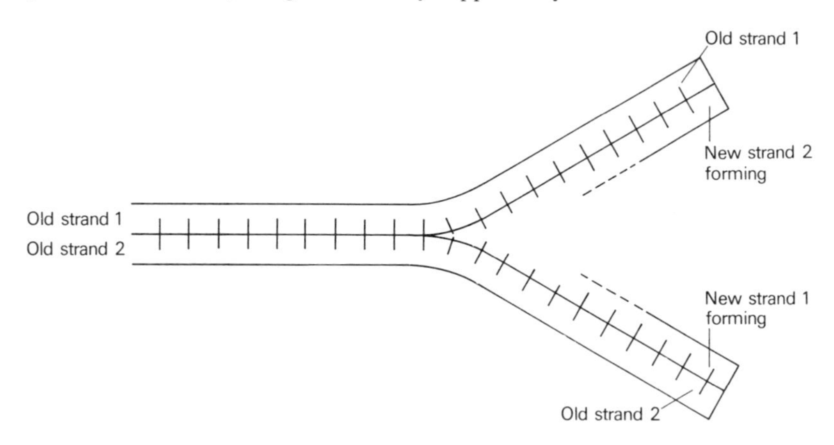

Fig. 8.3. Diagram illustrating separation of the DNA strands of a chromosome during cell division, with formation of new duplicate strands on the templates of the former pair.

Each strand is thus a counterpart of the other, in the sense that the sequence of bases on one strand indicates or determines the sequence on the other. When the two strands separate prior to cell division, therefore, strand 1 can act as a template for forming a new strand 2, and the old strand 2 for a new strand 1 (Fig. 8.3). In this way, bases are assembled on the new strands with the same, chemically determined, arrangement as before, namely that adenines will always be opposite thymines in the new pairs, and guanines opposite cytosines. The two daughter cells thus acquire either strand 1 and an exact copy of strand 2, or strand 2 and copy of strand 1. In this way every cell ordinarily contains a faithful copy of the DNA molecules of the cell from which it was derived, with the same sequence of bases as before. The sequence of the four bases in these molecules constitutes the code which tells the cell what to do, and what chemical substances to produce. The molecule is the message.

It would, however, be a poor vocabulary which consisted of four words only, and it is now clear that the sequence of bases has to be 'read' in groups of three, each such triplet of bases constituting a different word in the genetic code. This makes a vocabulary of 64 words (because the first base of each triplet may be any one of the four possible types, as may the second and the third, giving $4 \times 4 \times 4$ variants).

This is more than enough, because the function of the vocabulary is, essentially, to spell out the way to construct all the different proteins which the body contains and uses. Each protein consists of a long chain of amino-acids, and the body ordinarily uses only 20 different kinds of amino-acid in its proteins. A vocabulary of 64 words is therefore fully adequate to code for each amino-acid and indicate the sequence in which they must be assembled in synthesizing the different proteins for which the various chromosome pairs are responsible. Indeed, several amino-acids are coded for by as many as 6 different triplets, and only three are indicated merely by a single triplet. This lavish allotment of different triplets to code for the various amino-acids, which undoubtedly is saying something about our far evolutionary past, still allows spare triplets as punctuating instructions, for example, to say 'Start reading from here' or 'Stop. Protein completed', and so to indicate where to start reading each triplet. This is crucial, as any cryptographer knows, since a cryptogram written in blocks of three can make a very high grade of nonsense if begun at the wrong point (an dan yoneca nse ewh y).

We therefore have a basis for reading off the recipe required for constructing a complex protein out of the right sequence of the available amino-acids. The cell has its own procedure, of reading the recipe in the library of its nucleus, and making the protein in the kitchens of its cytoplasm. It copies the sequence of bases recorded on the chromosomal DNA (deoxyribonucleic acid) by forming molecules of the chemically rather similar RNA (ribonucleic acid), using the DNA as a template so that the sequence of bases on the RNA forms a counterpart of that on the DNA.

The strands of this 'messenger' RNA then migrate out of the cell nucleus, and carry the copy of the recipe into the ribosomes in the cytoplasm of the cell. The various ingredient amino-acids arrive, each kind being carried on its own specific vehicle in a manner reminiscent of the traditional Hindu pantheon, except that here the 'vehicles' are short sections of RNA. These molecules of 'soluble' RNA each engage with the appropriate triplet on the recipe in the ribosome, so that the various amino-acids are literally lined up in the correct sequence, to correspond with the sequence of bases in the original chromosomal code.

A typical protein consists of a few hundred amino-acids linked in series along its length. A gene programming the formation of that protein must therefore contain at least three times that number of bases along its strand of DNA. The 'at least' is required, because the sequence of bases in the DNA is known to contain long sections which do not code for parts of any protein molecule. (These 'intron' sections in the messenger RNA are, somehow, recognized as such, and excised from the recipe before it is used in the ribosome, a procedure which sounds closer to *Alice in Wonderland* than to formal biochemistry.)

The distance between successive nucleotide bases is about one third of a millionth of a millimetre (0.34 nanometres, as measured crystallographically). The length of DNA strand needed to code for a typical protein must thus be a few tenths of a micrometre, as measured along the strand, or a few micrometres (thousands of a millimetre) allowing for the non-coding sections of the gene. Since the total length of DNA in every cell is over one metre, there is evidently ample space in the total chromosomal library to code for each of the very many different kinds of protein that body cells ever produce, and to provide also for the additional genes that are needed to switch on and off the production of particular proteins according to the activity of the cell at the time and the organ in which it is situated. (In computer terminology, the fertilized ovum must transmit to each body cell at least 10 megabytes of information, taking account of the potential number of triplets, their redundancy in coding for amino-acids, and the non-coding DNA sections and strands.)

Since each pair of chromosomes contains genetic information from both parents, there will always (except on the sex chromosomes) be two genes that are potentially capable of giving instructions about each characteristic of cellular activity. There is no problem when both genes give the same instruction to the cell, for example to form proteins on the surface of the red blood cells which characterize those of blood group A. When, however, the gene from one parent instructs the cell to do so and the other instructs it not to do so, one instruction is ordinarily decisive and the other is ignored: in conventional terms, one gene is dominant and the other is recessive in its effect. In the same terminology, if a new mutation occurs which alters the normal behaviour of a gene, the mutated gene may be dominant in its effect in the child who inherits it from a parent, overriding the normal response to the unmutated gene received from the other parent. Alternatively, after a 'recessive mutation', the mutated recessive gene will not cause abnormality in the child if a normal gene is received from the other parent. A child who inherits genes of the same kind from each parent is said to be homozygous for that gene; if they are of different kinds he is heterozygous. A recessive characteristic is therefore only expressed in people who are homozygous for the recessive gene. A dominant characteristic is expressed even in those who are heterozygous for the dominant gene (Fig. 8.4).

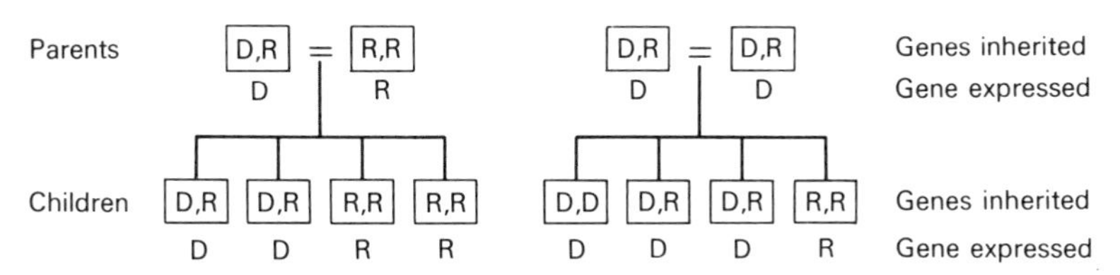

Fig. 8.4. Diagram illustrating the expression of recessive (R) characteristics only in homozygous individuals (with R, R genes) when recessive genes are inherited from both parents; or dominant (D) characteristics being expressed in individuals who are heterozygous (with D, R genes) or homozygous (with D, D genes).

In general, all new mutations are harmful. If the mutated form of any gene was appreciably more beneficial than the usual form in its effects, it would, over time, have been favoured in selection and would have become the usual form – the race keeps the aces that it picks up and discards the twos and threes; and the game has been going on for a long time.

Likewise, to the extent that they are harmful, mutated genes are progressively eliminated from the population – quickly for dominants and more slowly for recessives which are only expressed when they find themselves paired with an identical recessive gene homozygously and reveal their harmfulness. The frequency in the population of the various inherited defects or diseases which depend on mutation depends on the balance between the rate of new mutations of the relevant kind, and the rate at which their disadvantages, for example in reducing survival or fertility, cause their elimination.

For many inherited conditions, the situation is more complex than this, since the action of a number of genes may contribute to a single characteristic such as height or intelligence. In other cases, even though only one gene affects a particular body function, the effect may not be a simple alternative between dominant and recessive conditions. For example, for a gene determining the production of an enzyme, the concentration of the enzyme will be normal in people who are homozygous for the normal form of the gene, and much reduced in those who are homozygous for a mutated gene. It may however be slightly lower than normal in heterozygous people in whom a normal and a mutated gene are operating together, as if a single normal gene is not fully able to do the work of two. The separation between purely dominant and purely recessive characteristics is not as clear cut as it was for the Abbé Mendel and his peas.

In most species it is possible to distinguish between the different pairs of chromosomes, either by their shapes and lengths, or by the position of transverse bands of material which stains densely with certain dyes (Fig. 8.1). In simple organisms, which have only few chromosomes – the fruit fly has only four pairs – it has also been possible to find out which chromosome pairs carry the genes for various different inherited characteristics, and even the position within the chromosomes at which these genes are located.

These chromosome maps are based essentially upon knowledge of the abnormalities which occur together when chromosomes are seen to have broken or to have divided incorrectly. Abnormalities which are commonly associated in this way are likely to correspond to genes which lie close to each other on the chromosome, and which may become lost or misplaced together during cell division as a result of chromosome breaks. By similar methods, and particularly by the use of band staining techniques on cultured cells from individuals with inherited abnormalities, the human chromosome map is developing rapidly.

We can therefore picture the very large number of human genes as lying at definable positions along the length of the 23 pairs of chromosomes, like the knots on the coloured strings of the quipus carried by Inca messengers (and it would have been creepy if the quipu had been found to have 46 strings).

In a system as complex in its chromosomal divisions and in the transmission of detailed information from parent cell to daughter cell, it would be surprising if errors did not occur in this transmission of information. Equally, however, in a system which has formed the common basis for the heredity of all living species for all time, it would be surprising if there were not highly efficient arrangements for correcting most of these errors. The review of the genetic effects of radiation, as well as of many other mutagenic environmental agents, is a review of those forms of damage to DNA which may elude the efficiency of these cellular processes of repair.

TYPES OF MOLECULAR DAMAGE AND REPAIR

In some cases, the chemical problem is simple. If the 'backbone' of the DNA molecule is broken, by rupture of the bond between a sugar and an adjacent phosphoryl component (Fig. 8.2), there is evidence that the bond is rapidly reformed and that no harm results, provided that no other molecular damage has occurred. The scaffolding, of a histone protein, which holds the DNA helix in position, will ordinarily hold the cut ends of the DNA break together for long enough for this simple repair to occur.

Similarly if the bond is broken between a base and the sugar to which it is attached, a new base becomes attached, using the counterpart base on the second strand of the helix to ensure that the new base is of the appropriate kind.

The job is more complex when the damage involves the chemical structure of the base itself, or of the base and the adjacent stretch of the backbone of the strand. It is now apparently found necessary to remove and replace a stretch of the strand which may be longer by many base-to-sugar groups ('nucleotides') than the damaged section alone. These 'excision repairs' entail enzymatic teamwork of a remarkably high order. (It is difficult not to write in anthropocentric terms in discussing the subtleties of molecular biology.) An endonuclease enzyme identifies the position of the damaged section and cuts the molecular chain at a point beyond it. An exonuclease removes the damaged sequence and adjacent parts of the chain. A polymerase then organizes the synthesis of new bases in their correct sequence and positions, using the template of the intact second strand. And finally a ligase ties the job up at both ends.

Thanks, perhaps, to this meticulous demarcation of duties, the repair is completed rapidly, within hours, and is ordinarily error-free. The speed of the repair has been estimated by measuring the rate at which broken fragments of DNA strands become rejoined after irradiation. Soon after irradiation the DNA extracted from cells contains fractions which sediment unusually slowly through fluid (an alkaline sucrose solution), showing that these fractions contain shorter lengths of DNA than normal. In DNA extracted from cells at rather longer times after irradiation, the proportion of these shorter fragments decreases progressively until all DNA is again of the normal length, indicating the speed with which broken strands are rejoined.

This evidence of rapid repair of single strand breaks is consistent with the radiobiological evidence that two radiation exposures usually cause less harm if they are separated by a long time interval than by a shorter one. In the longer interval, there has been time for most of the single strand breaks to be fully repaired. In the shorter interval, many remain unrepaired and the second exposure may cause breaks on the second strand close to unrepaired breaks on the first strand; and double strand breaks carry a greater chance of being incorrectly repaired and so resulting in permanent damage.

When both strands of the helix are broken at about the same level, neither can be repaired simply with the aid of the other as an intact template. While both strands may in fact be rejoined, this is liable to be at the expense of losing parts of the DNA sequence of each strand, with failure of some components of the functional activity of the cell if these parts include an active coding sequence. Alternatively, if the strands are not rejoined, the corresponding parts of the chromosome may fail to replicate or be distributed normally to daughter cells at the next cell division. In contrast to breaks of a single strand, therefore, double strand breaks, although less frequently caused by most forms of radiation, are likely to be more slowly and less correctly repaired, and so to involve greater risks of harm.

Much remains to be learnt about the molecular basis of radiation damage and its repair; and precisely which of the various methods of repairing DNA that have been identified in plants and bacteria are of importance in human or other mammalian cells. Enough has been established, however, to show that a number of types of damage and repair are common to many living species; that most of the damage caused by low doses of radiation and other mutagens is correctly repaired; and that hereditable defects result from the part of this damage which is incorrectly repaired but which nevertheless allows the cell to survive.

It seems clear also that what is worst for the cell is best for the species. Unrepaired breaks are likely to cause death of the cell so that no defect is transmitted. It is the breaks which are incorrectly repaired but allow the cell to survive which are transmitted as genetic defects. The death of a single germ cell would be undetectable. A damaged but surviving cell, however, can transmit harm of which the kind and severity depend on the kind and amount of genetic information which has been altered or lost.

Genetic defects are ordinarily described as resulting either from mutations or from chromosome aberrations. Mutations are attributed to local damage within a chromosome, but with preservation or repair of the continuity of the chromosome as a whole. Chromosome aberrations result from unrepaired breaks in the chromosome, so that fragments of a chromosome become lost during cell division, or incorrectly attached to other chromosomes.

In this sense the effects of radiation or other mutagens in causing mutations or chromosome aberrations merely represent different effects of the same basic types of damage, or differences in the mode or efficiency of its repair. Indeed, there is no clear dividing line between the two categories of chromosomal damage, and it is likely that many effects classed as due to mutations may be attributable to small chromosomal rearrangements which are not, or not yet, detectable microscopically as such, but which alter the manner or the regulation of gene function.

The difference in the category under which genetic effects are described does, however, have an important biological implication. Chromosome aberrations are detected by microscopic examination of cell nuclei, but individual aberrations recognized in this way may or may not be associated with any abnormality in the individual from whom the cells were derived. Mutations, on the other hand, are inferred to have occurred when an individual shows some abnormality, of metabolism or structure, which is believed to have resulted from changes arising *de novo* in the germ cells of a parent or an ancestor.

Both types of effect include a wide range in the severity of harm that they involve. In some types of chromosome aberration, the normal amount of chromosomal material is transmitted to subsequent generations, whereas in other types, whole sections of some chromosomes are lost in subsequent cell divisions. From many, no identifiable harm appears to result, but in some cases specific defects or diseases have been shown to be associated with loss of sections of particular chromosomes.

Similarly in some mutations, the defect consists only in changes in the colour of parts of the iris of the eye, or in biologically insignificant alterations in the blood concentration of some enzyme. In others, severe abnormalities may result, either as dominant conditions in the first generation or as recessive conditions expressed in later ones. Estimates of the genetic risks of radiation exposure need to take account of the number and the severity of the effects that are likely to be caused by different radiation doses.

ESTIMATION OF GENETIC RISKS OF RADIATION

Ample evidence has been obtained in various animal species, and particularly in the mouse, on the frequency with which radiation induces various types of genetic abnormality. In man, however, no reliable data have been obtained on which such estimates can be directly based. There are, indeed, no circumstances in which it has been possible to study the frequency of inherited abnormalities in the progeny of large populations of people who have been substantially irradiated, and to compare this frequency with that in a population which was similar in all essential respects except for not having been equally irradiated.

It might be thought that populations living in areas of high natural background radiation might usefully be compared in this way. Where the terrestrial radiation rates are high, the studied populations have hitherto been too small to show any effect unless it had been a large one. No such effect has been demonstrable, either in a careful recent study in Southern China,[61] or reliably in Southern India where the lack of comprehensive medical statistics makes accurate comparisons difficult.[62] Where radiation rates differ between communities because of differences in altitude, it is again hard to rely on simple comparisons, because populations in mountain and lowland areas may differ in factors such as the efficiency of registration of abnormal births, the post-natal survival rates or the degree of inbreeding, which could influence the recorded frequencies of inherited defects.

Even in the groups of considerable size exposed at known but differing doses in Hiroshima and Nagasaki, no clearly significant excess of inherited abnormalities has yet been detected on which a reliable genetic risk estimate for man could be based. In this instance, however, the largely negative findings among children conceived since the bombs are of value in indicating – as discussed below – that estimates based on results observed in the mouse are at least unlikely to underestimate the radiation risks in man.

The present necessity to place reliance on genetic risk estimates observed in mice and other species is, however, less precarious for inherited defects than it would be for cancer induction. In cancer induction it is known that various immunological and other body mechanisms can cause extensive destruction of the cells of a developing cancer; and the frequency with which cancers develop after equal irradiation of the same organs differs considerably in different species of animals or strains of the same species.

As regards genetic effects also, there are admittedly substantial differences between different animal species in the types of DNA damage and repair which are important genetically, as well as in the number and behaviour of the chromosomes. These differences, however, are considerably less between different primates or different mammals than they are over the whole range of species in which faultily repaired DNA appears to determine inherited abnormalities. The study of the genetic processes in a wide variety of animal and vegetable species has given us a considerable understanding of the linkage between radiation exposure and inherited damage in a qualitative sense. As a basis for quantitative estimation of radiation risks in man, however, reliance can best be placed on the extensive studies that have been made on the mouse and some other rodents, supplemented by more limited findings in primates, and on human cells cultured *in vitro*.

In this connection it is important to note that important chromosome aberrations are induced in human lymphocytes with the same frequency per unit dose whether such cells have been exposed in the body in the course of radiotherapy, or exposed to an equal dose *in vitro*[63, 64] (the frequency of dicentric chromosomes being about 2 per cent per sievert in each case). The frequency of such chromosomal aberrations gives no indication of the risk of any induced disease or form of inherited harm. However the equal rates of induction when these cells are irradiated inside or out of the body does give a clear indication that no immunological or other general body reaction appears to intervene in man between the primary radiation damage to DNA and its expression in chromosomal changes, of which a proportion results in inherited abnormalities. In addition, it strengthens the likelihood that comparisons between the sensitivity of human and other primate or mammalian cells irradiated *in vitro* should give useful guidance as to the relative sensitivity of the cells of man and of other species when irradiated within the body, at least as regards the types of cells studied. Because of the essential similarities between the effects of radiation on DNA in different mammalian cells, the use of animal data appears to have reasonable validity as a starting point for the quantitative estimation of genetic risks in man.

Two largely independent methods have been used to infer the probable genetic risk in man from that observed in the mouse or other species.

'Doubling doses'

The first method rests on the remarkable observation that about the same dose of radiation is needed to double the natural frequency of mutation responsible for each of quite a large number of separate types of inherited abnormality in the mouse. This broad similarity in size of doubling dose applies to different methods of inheritance, for example through chromosomal aberrations or point mutations, or by induction of dominant or recessive mutations, or in effects on male or female germ cells. A dose of about 1.5 sieverts, given protractedly over a period of time, was required to double the natural frequency of abnormalities inherited in these ways, as judged by the results of 18 separate studies of the effects of radiation on different genetic processes.[65] Substantial variations in individual estimates of doubling dose are to be expected, owing to inaccuracies in its experimental determination. The variations are, however, no greater between different modes of inheritance than they are for different investigations on one such mode, and the doubling doses in 14 of the available studies lay between 0.7 and 2.4 Sv. If hereditary defects are normally caused by a variety of mutagenic agents, including radiation, that are present in the environment, and if all these agents act by causing similar changes in the structure of the nuclear DNA which are repaired by the same cellular mechanisms, then an added exposure to any one of these mutagens should indeed increase proportionately the frequency of each type of genetic defect which is induced by normal environmental agents.

Whatever the explanation, however, the observations suggest that the genetic risks in man from radiation might be inferred similarly from the normal frequency of inherited human abnormalities, provided that two pieces of information were available. First, we would need to know the amount of radiation which doubled the natural frequency of at least one type of inherited abnormality in man; and, secondly, we would need an estimate of the proportions of different inherited abnormalities which were due to past mutations, rather than to any selective advantage which maintained the frequency of these abnormalities.

No direct estimates of genetic risk have been obtainable in man, and no value can therefore be given with confidence for a human doubling dose. The absence of any clearly detectable excess of inherited abnormalities in children born since the atomic bombing of Hiroshima and Nagasaki, however, appears to exclude a doubling dose any lower than that observed in the mouse. Moreover several observations in these children do suggest a slight increase in frequency of genetic defects of several types, although the increases are individually too low to be of statistical significance. Thus there is a suggestion of some increase in frequency of stillbirths, of defects causing death in the first week after birth, and of certain kinds of chromosomal abnormality. The frequency of these effects would suggest a doubling dose of about 1.6 Sv to germ cells for the exposure that their parents received from the bombs.[66] This tentative estimate of a human doubling dose is therefore about equal to that of 1.5 Sv estimated in the mouse as the average for various genetic effects of protracted exposure. It at least suggests that the sensitivity of human germ cells to radiation is no greater than that in the mouse.

In deriving estimates of the genetic risk of radiation in man, UNSCEAR has assumed a rather lower doubling dose of 1 Sv. This degree of caution reflects the uncertainty of the evidence that human germ cells are no more sensitive than those of the mouse. The UN committee, in its 1982 report to the General Assembly, also estimate that 1.5 per cent of all liveborn children normally had significant inherited abnormalities of which the frequency was likely to be increased in this way. If then a dose of 1 Sv per generation would eventually double this natural frequency, the risk of abnormality resulting from parental exposure to radiation would be of 1.5 per cent per sievert; or, assuming a proportionality between dose and frequency of effects, a risk of 1.5 abnormalities per 100 000 descendants of parents exposed to one millisievert.

This makes it possible to assess the likely importance of environmental radiation in contributing to genetic abnormalities, on the basis of the estimated risk of parental exposures. By the age at which parents conceive children, at present averaging a little less than 30 years, a parent will have received about 30 mSv of gonadal irradiation from natural radiation sources. For a risk of 1.5 abnormalities per 100 000 children per millisievert exposure of their parents, natural radiation sources would cause 45 abnormalities in every 100 000 children, as compared with the 1500 per 100 000 occurring naturally and regarded as attributable to mutation. To this value of 45 from natural sources of radiation, medical irradiation would add about 6, and all other present sources of radiation exposure would add a further 1 if continuing at present rates.

Direct method

A second, and more direct, method of assessing the genetic risks of radiation depends upon separate estimates that can be made of the sensitivity of the male and female germ cells in various species for the induction of dominant or recessive mutations in the sex or other chromosomes, and for the development of genetically important chromosomal aberrations. For each of these types of effect, use is made of the best available quantitative evidence in any species, with proper allowance for the natural contribution of each such effect to the total of inherited disabilities in that species and in man.

The frequency with which radiation causes chromosomal aberrations can be more easily studied than the frequency with which mutations are induced, since abnormalities of the chromosomes are detectable microscopically after irradiation, whereas the scoring of mutations depends upon maintaining breeding colonies of animals and comparing the small numbers of abnormal progeny in irradiated and unirradiated groups. For this reason there is fuller information on the frequency with which chromosomal aberrations are induced in a number of mammalian species, including primates, than has been obtained for mutations. There is also a good understanding as to which types of chromosomal abnormality are likely to allow the cell to survive and divide, but to involve a loss of chromosome material in a proportion of the daughter cells at cell division. Few detailed links have been established between abnormalities of particular chromosomes and any specific inherited defects that they may cause. A numerical association can be made, however, between certain types of aberration, and the probability of resultant inherited defects.

On these bases, estimates can be made of the probability of an inherited defect in a child of an irradiated parent; or, with less confidence by this method, in all of his or her descendants. In each case, the probabilities are higher following irradiation of the father than of the mother. As an average value, however, the frequency of such defects in the first generation is estimated to be between 0.5 and 2.1 per million children per millisievert received by the parent. This estimate by the direct method is thus a little lower than that of 2.2 per million per mSv in the first generation as assessed by the doubling dose method.

In summary, we must assume that radiation exposure increases the normal frequency with which mutations and chromosome aberrations occur, and that such effects are in general to a greater or less degree harmful. The probability of an inherited abnormality is greater after a given paternal exposure than after an equal maternal one. As an average value, however, a single dose of 1 mSv to a parent gives a 2 in a million risk of any substantial defect in his or her child conceived subsequently; or a 15 in a million chance of any such defect resulting in all subsequent generations.

It should be emphasized that these are the estimated risks of parental exposure: that if a million *parents* or prospective parents each receive 1 mSv, 15 substantial defects will result. If a million *people*, of all ages, each receive 1 mSv, the expectation would be of 6 resulting defects, since only about 40 per cent of the people exposed would, in the UK, be at ages at which they will subsequently conceive children. (In England and Wales in 1977, the average ages at which people conceived children were about 29 in men and 26 in women. The average ages at which people died were 69 in men and 75 in women. The proportion of people in whom irradiation could cause inherited effects was therefore 0.42 in men and 0.34 in women, as noted on p. 48).

9

CELL TRANSFORMATION AND THE INDUCTION OF CANCER

A great deal is now known about many of the causes of cancer.[67] Little is known with certainty, however, about the way in which individual body cells may become 'transformed', so that they divide repeatedly in an uncontrolled fashion, and ultimately form a recognizable cancer. For our purposes, several stages in this process are important.

CELL TRANSFORMATION

Whatever its cause, the original cell transformation must be such that the cell will transmit to its daughter cells, and they to their daughters, the same capacity for unlimited cell division. This clearly suggests that a primary change must be a mutation in the gene structure of the cell chromosomes, since it is by this means that information is transmitted to daughter cells when the cell divides. It is in this sense that cancer induction by radiation may result from a gene or chromosomal change in cells of one of the somatic body organs, just as hereditarily transmitted effects are due to genetic or chromosomal damage in a cell of the germinal tissues, even though the type of causative mutation will differ.

In the case of somatic cells, an abnormality of metabolic behaviour of a single cell, or of the daughters of a single cell, would not be likely to cause any significant abnormality in the function of the organ in which it was situated, unless the change in its behaviour was one which gave it a selective advantage in multiplication over the other cells of the organ. Thus a mutation in one cell amongst the millions of cells in an organ, which reduced or changed its enzyme or other protein production, could never in itself by detectable. The mutation would be of great importance, however, if for example it altered the protein structure of the surface membrane of the cell, so that the cell was not 'recognized' and dealt with efficiently by the processes which normally restrict cell division to the rate appropriate to the organ or situation in which the cell occurs.

Both the inherited and the carcinogenic effects of radiation may thus be due to mutation in single cells which subsequently multiply in such a way that their changed behaviour becomes manifest and biologically important.

INITIATION AND PROMOTION

A large number of chemical substances and physical agents have been found to influence cancer development. Of these, some are effective in causing the initial transformation of a normal cell into one with the potential for repeated division. They do not themselves, however, give the stimulus for this cell division. Other agents do not cause such transformation, but do promote the repeated division of cells which have already been transformed. Yet other agents, of which radiation is one, appear to have both initiating and promoting actions. In discussing radiation carcinogenesis, therefore, we need not be greatly concerned with the distinction between these two kinds of effect. They may be significant, however, in our understanding of the long intervals that elapse between exposure to radiation and the appearance of any resulting cancer.

LATENT PERIOD

The length of this period of latency is identifiable when a brief radiation exposure, occurring at a known time, has been followed by a subsequent increase in the number of cancers developing in the exposed population, in excess of the numbers to be expected naturally. These conditions applied in Hiroshima and Nagasaki, where the frequency of cancer development has been recorded in a large group of exposed people over many years. In these circumstances the mortality rate from leukaemia was higher than that expected in unirradiated people from within a few years of the bombing until about 30 years after it, by which time the rate had returned almost to a normal level. These rates are based on the number of people still living at the relevant times, and therefore indicate a true limit to the length of latency, and not simply a decrease in surviving numbers of the population studied.

The average length of this interval between the radiation exposure and the deaths from leukaemia attributable to it, was about 14 years; and the interval between exposure and onset of the disease averaged about 12 years. Other surveys have identified the mortality from leukaemia as being raised during the period from 5 to 20 years after certain forms of radiotherapy.[29]

For most other forms of cancer, the average latent period between exposure and the appearance of the cancer is even longer than for leukaemia. This average interval certainly exceeds 20 to 25 years, but is difficult to estimate with certainty for several reasons. For one thing, it is not yet clear whether the increased frequency of onset of such cancers reaches a maximum value and then decreases to zero, as with leukaemia, or continues for the remaining lifetime of the exposed people, since no surveys have continued for long enough to answer this question. Moreover, in at least some types of cancer, the average length of latency has been found to vary according to the age at exposure, so that no simple averaging of interval is appropriate. And even in the Japanese A-bomb survivors, the cancer rates are in most cases insufficiently raised above the expected natural rate to allow a clear indication of the time course of the increase. It seems likely, however, that few solid cancers develop within 5 years of exposure, and that half of those that have been induced will only develop after 20 or 25 years.

Contributions to latency

At least three factors may contribute to these long periods before cancers appear. First, there may be a substantial interval between the action of the radiation as a cancer initiating agent, and its action, or that of any other environmental agent, as a promoter of cell division. If so, however, the magnitude of any such interval is not known. Secondly, even if all the dividing cells of a developing cancer were to survive, many generations of cell division would be needed before a single transformed cell could multiply to produce a tumour mass that could cause any symptoms or effects, or be detectable by clinical means. For example, the increase from a cell of volume 100 cubic micrometres to a tumour of volume one cubic centimetre involves a factor of 10^{10} increase in size, which is equivalent to about 33 doublings of cell mass by successive divisions. And thirdly, and almost certainly as the major component, there is ample evidence that most of the daughter cells in a dividing cancer clone do not in fact survive. Indeed, in certain types of apparently radiation induced cancers, the so-called microsclerosing papillary tumours of the thyroid, for example, the tumour typically does not increase beyond a few millimetres in diameter during a lifetime, and can be classed as a 'cancer' only on account of its microscopic appearance, and not in terms of any malignancy of behaviour in the body.

One of the factors which has been regarded as having a promoting effect on tumour development has been the presence of severe damage to tissues and the subsequent distortion of their normal structure with the growth of fibrous tissue. When, therefore, cancers were found to follow high radiation doses which caused such tissue damage, it was at first thought that the severe fibrosis might have caused the cancers to develop.

This view was the more plausible, since it is now clear that cancers would occur only rarely in tissues that had been exposed to doses below those causing fibrosis – being induced in only a few per cent of cases at most, and then commonly only at more than 20 years after exposure. Without careful epidemiological studies, extending for many years after irradiation, and comparing the cancer incidence in quite large numbers of exposed people with that in the unexposed, no increase in cancer frequency would have been detectable.

EPIDEMIOLOGICAL STUDIES[29]

Some of the earliest such studies were based on the causes of death of radiologists in the USA who died in the years following 1938, and who would in many cases have been occupationally exposed to radiation at quite high doses from the earliest days of radiological practice. These investigations showed that the mortality rate from leukaemia was significantly greater in these than in other medical specialists who did not work with radiation, and that leukaemia was being induced by amounts of radiation which were certainly too low to cause gross tissue damage and fibrosis. They did not, however, show how low a dose was having this effect, since the exposures that these early radiologists had received had not been measured, at least during most of their working lives, and could not be estimated with any accuracy.

Since then, however, a large number of surveys have been made in groups of people who had been exposed to known doses under various circumstances, and in whom the occurrence of any increase in cancer during the following decades could be detected.[22,29,86] In this way a considerable body of numerical information has been built up, not only of the amounts of exposure which have been followed by an excess incidence of cancers, but also of the way in which the size of the excess varies with the size of the radiation doses to which it is attributable.

Such information is obviously essential to any assessment of the risks of radiation exposure in general, and also of the risks from particular sources or types of exposure. The collection and interpretation of this information however involve a number of problems. It is necessary to evaluate the cancer risk, not only of irradiating the whole body at a given dose, but also of irradiating the various body organs individually, or at least those organs which may be selectively irradiated by the concentration and retention of different radionuclides in them. The whole body is irradiated more or less uniformly by many types of external exposures, and by internal exposure from various radionuclides such as radioactive caesium, carbon, potassium, or hydrogen when they are taken into the body in a chemical form which becomes generally distributed through all body tissues and fluids. On the other hand, radioactive iodine usually becomes most highly concentrated in the thyroid gland; inhaled plutonium irradiates first the lung, and later the liver and the bones, in which it is retained; and insoluble compounds of radionuclides, if swallowed, cause irradiation which is mainly limited to the lining of the stomach and the intestines.

The cancer risk of radiation exposure, either of the whole body or of individual organs, can be estimated reliably if the investigation meets a number of conditions:

1. The radiation doses which the body or the organs have received need to be known or adequately estimated. Ideally, the exposures should cover a range of doses, so that the variation of cancer rate with dose can be determined – the presence of such variation giving evidence that any excess of cancers is due to the amount of radiation exposure and not to some other, associated, cancer-producing agent.

2. The whole of the exposed group, or at least the great majority of it, should be followed for a long period: ideally for the remainder of their lives,

but in any event for over 20 years, since it cannot be assumed that an increased cancer rate observed over shorter periods will subsequently continue or will decrease. It is also important to distinguish, by such prolonged surveys, between the causation of an increased number of cancers, and causing the normal number of cancers to develop earlier in life than they would otherwise do.

3. This follow-up should ensure efficient ascertainment of the causes of all deaths, as based if possible upon death certificate records. The reliability even of such certificates varies considerably. Cancer may well be recorded as the cause of death, but without correct identification of the organ in which the cancer arose, unless autopsy or biopsy findings are available. This distinction becomes important in estimating the risk that cancer may be induced in specific organs when they are selectively irradiated.

4. The number of cancers which develop, or which cause death, in the exposed group needs to be compared with the number observed or expected in a group which is identical in all relevant respects except that they have not received the additional radiation exposures. This requirement is in general the hardest to achieve. Ideally, the comparison group must match the exposed group in many ways: in sex and age distribution; in both length and calendar period of follow-up; in the presence of any disease for which the radiation was given, and in the severity of such disease; and in various factors such as genetic composition, 'social class', occupation and region of residence, since some cancer rates are known to correlate with such factors. The importance of this correspondence between the exposed and the comparison group is considerable in studies at low doses, since even very small differences in numbers of cases, due to mismatching of the two populations that are compared, could be misinterpreted as being due to a large excess of cancers induced per unit radiation dose.

5. It is a further essential requirement that the size of both the exposed and the comparison group should be large enough to allow small increases in cancer rates to be detected with statistical significance. The need here varies with a number of factors. A small increase in the incidence of a cancer will be more easily, and more reliably, detected if that type of cancer is normally rare than if it is common; and much larger groups will need to be examined in the latter case if the size of the increase is to be established with confidence. Similarly the detection of any radiation effect at low dose is more difficult, and requires larger groups to be studied, than is the case at higher dose. The statistics here are relentless. If an excess number of cancers due to radiation can be reliably estimated by study of 1000 people who had been exposed to 1 sievert, it would require the study of 100 000 people exposed to 0.1 sievert to estimate the excess with equal reliability, if the induction rate per sievert were the same.

Given all these problems, it is perhaps remarkable how much valid, and reasonably consistent, information has been obtained on the cancer risks of radiation in different body organs, and from a variety of different sources. Each of these sources has its own defects. Together, however, they allow a reasonably firm estimate of the risks of various forms of radiation exposure at moderate dose, and an approximate inference of the likely risks at lower doses.

EVIDENCE FROM OCCUPATIONAL EXPOSURE

Luminizers

The use of radium in luminizing the dials of watches and other instruments gave rise to two types of radiation hazard, in the early days before the importance of any hazard from moderate exposures was recognized. The major risk resulted from an intake of radium into the body, when the brushes used to apply the paint were often licked to give them a fine painting point. A lesser risk, more recently recognized, was due to external irradiation of the body from the supply of paint that was kept on the working desks.

Radium becomes concentrated and retained in bone, as a result of its chemical similarity to the calcium of bone. In the days before artificially produced radionuclides became available, the long-lived isotope radium-226 was commonly used to give a continuing luminescence to the paint. When any of this radium is taken into the body, however, its incorporation in bone causes prolonged irradiation of bone cells and of structures that are immediately adjacent to bone surfaces.

As early as 1924, an increased number of deaths from osteosarcoma, the cancer of bone cells, was seen to be occurring in those who worked, or had worked, in the luminizing industry. A few years later, an increase was detected also in cancers of the cranial antra and sinuses, the air-filled cavities within the skull of which the walls are formed by membranes whose cells lie immediately adjacent to the bones of the skull.

Extensive studies have therefore been made on the number of these cancers which have occurred, as compared with the numbers expected in populations of the same sex and age distribution but unexposed to radium. The excess number of these cancers, per thousand people exposed, then needed to be related to the amounts of radium retained in the bones.

Because of the long half-life of radium-226 and its long retention in the body, it was possible to measure directly, by whole body counting methods, the average amounts of radium retained in groups of luminizers, and so to infer approximately the amounts taken into the body at earlier periods, and hence the total radiation dose that bone cells would have received. In some instances also, the actual concentration of radium in bone has been measured in samples of bone removed after death, whether from bone cancer or from other causes. In these ways, and since radium retained in the body is located largely in bone, reliable information has been obtained on the risks of inducing bone cancer and the corresponding doses that bone cells had received.

This has not been possible, with equal confidence, for cells of the cranial antra and sinuses, since the radioactive gas radon, released during the decay of radium, accumulates to an unknown extent in the air of these cavities, instead of being rapidly removed from bone surfaces by the blood circulating round them. The membranes of these cavities will thus be irradiated both by the radium in bone and by this radon and its daughter products, the dose from radium only therefore giving a lower limit to the total dose that they will have received.

The frequency with which bone cancer was induced by known amounts of radiation was first reported in 1930 – many years before epidemiological surveying had shown any other form of cancer to be caused by such moderate radiation exposures. There is a reason for this priority. The high concentrations of radioactive material in bone caused cancers to develop rather frequently in the bone cells in which, under normal circumstances, cancer is rare. In bluntly engineering terms, the 'signal to noise ratio' was high. The effect of radiation was therefore identifiable, and measurable, in a much smaller group of exposed people than if similar radiation doses had been delivered to tissues in which the natural cancer rate had been higher.

Partly, perhaps, for this reason, it was only much later that a lesser effect of external radiation of the body from the contents of the paint pots was identified, with a significant increase in breast cancer reported in 1980 in women who had worked as luminizers before the importance of shielding the radioactive sources was fully recognized. The size of the radiation risks to breast tissues, as a probability of cancer induction per sievert, can only be very approximately estimated, by tentative reconstructions of the early working conditions. These estimates appear consistent, however, with those based on other surveys.

Uranium and other hard-rock miners

Long before the earliest use of luminous paints, it was known that the pitchblende miners of the Schneeberg and Joachimstal districts died of a lung disease at an alarmingly high rate.[26] With increasing medical knowledge, certain forms of lung cancer were recognized as responsible for most of this increased death rate, but the cause of these cancers was disputed for a long time. It became clear, however, that the rock in these and in some other hard-rock mines, although not ordinarily in coal mines, contained substantial concentrations of uranium. As a result, in the absence of adequate ventilation and other precautions, considerable radiation exposure of miners was occurring in two ways. Firstly, external exposures came from the walls of underground shafts and galleries, largely from the accumulated decay products of the uranium, and particularly from radium-226 retained within the rock. And secondly, and more importantly, this radium continuously released its gaseous daughter radon-222. When inhaled, this gas, and its own daughter products, irradiated the linings of the bronchial passages and the air sacs of the lung.

With modern techniques of ventilation, washing of working rock faces and sealing off of disused shafts, the radon concentrations in working areas can be greatly reduced.[68] Today's death rates from lung cancer still reflect the working conditions of several decades ago, however, owing to the long latencies of cancer development. Many studies during the last 10 years in fact still show increased death rates from bronchial and lung cancers, not only in uranium miners in Czechoslovakia, the USA, Canada, and China, but also in other hard-rock miners in Sweden and the United Kingdom, where radon concentrations in underground working areas are often substantial.

It is difficult to establish the relationship between increased cancer rates and the radiation dose to lung tissues accurately for several reasons. First, the amounts of radon inhaled during a working life can usually only be assessed on the basis of air sampling devices placed at various positions in the mines in which the men have worked, although the validity of these estimates has been checked in a few cases by using sampling equipment actually worn by miners in the course of their work. Secondly, the dose that will result from a given air concentration of radon depends on a variety of factors such as the breathing rate and the size and concentration of dust particles in the air, since radon daughters are adsorbed onto these particles. The dose in any case differs for different parts of the bronchi and lungs from which cancers may originate. Thirdly, other factors in the working environment, such as diesel fumes or the dust itself, may influence the lung cancer rate. The comparison of the cancer rates with those expected in non-miners is, moreover, rendered unusually difficult by the importance of cigarette smoking as a major cause of lung cancer, and the uncertainties in matching the smoking habits of miners with those in the unirradiated population with which they are compared. Finally, the dose to lung tissue by external gamma radiation from the walls of shafts will add to that delivered internally by alpha radiation from the radon daughter products, and require additional estimates of the exposures received, even though these contributions to the total dose will usually not have been major ones.

Despite these difficulties, broadly similar estimates have been obtained in six countries of the risk of lung cancer after inhaling radon in mines, and of the way in which the risk increases with the amount likely to have been inhaled, and the resulting dose of alpha radiation to the lungs from the radon daughter products. It is important to have a reliable estimate of this risk, since a major source of exposure following inhalation of plutonium in insoluble form is also by alpha radiation of the lungs.

With one exception, no evidence has been obtained of an increased frequency of any other form of cancer in these mines; and no appreciable increase is likely, since the inhaled radioactivity remains largely confined to the lung, and since the doses from external radiation are not large. One survey, however, suggests an increase in cancers of the skin of the face in Czechoslovak miners. Insofar as non-mining populations form a valid comparison group for miners, who must experience various irritants of the exposed skin, these data would be consistent with an effect of radiation in any areas of skin which were thin enough for the short penetration of alpha rays to reach layers of living cells lying below the skin surface. No estimate is possible, however, of the doses which such cells might receive.

EVIDENCE FROM RADIATION THERAPY

Radiotherapy of ankylosing spondylitis

One of the earliest and most valuable assessments of the carcinogenic effects of radiation was based on studies in 1956 and 1965 of the causes of death of patients who had had radiotherapy for ankylosing spondylitis, a crippling form of spinal arthritis which could often be considerably relieved by irradiation of the most affected parts of the spine. The causes of death were identified in about 1600 people who had been so treated during a period of up to 25 years previously. It was found that the number of deaths from cancer was slightly greater than would be expected, in the same period and during the same calendar years, in a normal population with the same distribution of sex and age, as judged from national statistics. It was further found that the increase was in types of cancer arising in organs lying within the irradiated areas of the body, and that there was no indication of any increase in cancers of those organs which would not have been irradiated. Deaths from leukaemia were increased, and this is consistent with the fact that this form of cancer results typically from proliferation of certain types of cell in the bone marrow, and that a proportion of the bone marrow would have been included in the areas irradiated.

The value of this study lay in the facts that many patients had been treated in this way, the doses of radiation and the areas to which they were delivered were known, and the causes of death recorded in the death certificates were traced in most of these cases. Moreover, despite the disability caused by the disease, most treated patients survive for long periods. In the majority of cases, therefore, there is adequate time for any harmful effects of radiation to be expressed, in spite of the long latency typical of radiation induced cancer.

One weakness of the original investigation was that patients with this disease might have an increased liability to develop cancer, whether treated by radiation or not. The comparison with cancer mortalities in the general population might therefore be invalid. It is indeed known that patients with ankylosing spondylitis are prone also to a disease of the colon, in which cancer of the colon sometimes occurs as a sequel. For this reason, such cancers were specifically excluded from the analysis. There might, however, still be other cancers of which the association had not been recognized, although it is unlikely that these should involve only the areas to which radiation was given in treatment.

This difficulty has since been removed by a parallel study of patients with ankylosing spondylitis who had not been treated by radiation. In these patients no excess of cancer occurred, as compared with the mortality observed in the general population. This evidence also served to exclude the possibility that the excess in irradiated patients was due, not to the radiation, but to drugs normally given in this condition for the relief of pain.

Meanwhile this study, and later extensions of it, appear to yield reliable and moderately precise estimates of the risks of radiation induction of cancers in a number of exposed organs after average doses in the region of a few sieverts; and also of leukaemia following exposure of part of the bone marrow, and of the colon, if reliance is placed on the lack of any increase of colonic cancers in the unirradiated patients.

Injection treatment of ankylosing spondylitis

The spinal irradiation that was found to be helpful in treating ankylosing spondylitis was formerly delivered by a different method in certain European clinics. Instead of external irradiation of affected areas of the spine by x rays, a short-lived isotope of radium was used to cause internal irradiation of bone surfaces. The radium-224 was injected intravenously and, as with the radium-226 in the dial painters, became concentrated in bone. Because of its short half-life, of only 3.6 days, it loses its radioactivity by decay before it has migrated into bone from the bone surface, onto which – like calcium – it is deposited initially. It therefore irradiated these surfaces, and the spinal joints lying adjacent to them, and it had a similar therapeutic effect to that of x rays, without a comparable exposure of other body organs.

Unfortunately, however, the cells responsible for the initiation of bone cancers are located close to the bone surface, and patients treated in this way were found to have a significantly increased mortality from such osteosarcomas, although not from cancers of other body tissues. It is a coincidence that the risk of death from the cancers caused by this treatment was about equal to that from all cancers induced by the external radiotherapy of this condition, involving about one per cent of patients treated by either method. Both types of treatment have now been largely abandoned, although it is for consideration whether, at the time when they were used and if equally effective alternative treatments were not available, this risk of death occurring many years later would or would not be thought to be justified in the effective relief of a painful and crippling disease.

Several large surveys of patients treated by radium-224 injection for ankylosing spondylitis and certain other diseases have yielded estimates for the risk of irradiating bone cells, and for the variation of this risk with age at injection, and with the period over which the course of injections was given. Taking account of the increased bone cancer rate, and the average dose to bone to which it was attributed, the values of risk for a given dose from the radium-224 differed from the estimate derived from the studies in the dial painters exposed to incorporated radium-226. When it was appreciated, however, that the bone cancers derive predominantly from cells on the surfaces of bone, this apparent discrepancy disappeared. The short-lived radium-224 delivers essentially all of its dose to bone surfaces. The much longer-lived radium-226 gradually becomes distributed throughout bone. When the bone cancer risk is calculated in terms of the cancer probability per unit dose to the relevant 'endosteal' cells on bone surfaces, and not to the dose as averaged throughout the whole mass of bone, the risk estimates derived from the two sources are in substantial agreement. Indeed, the values obtained after radium-224 exposure are in one way more reliable than those from radium-226. The latter isotope, owing to its long half-life, continues to irradiate bone cells during the whole latent period of any cancers which it may have caused. It is therefore difficult to be sure how much of the dose delivered by the time the cancer is detected, needs to be discounted because it was delivered after the cancer was originally initiated. With radium-224, however, this uncertainty does not arise, as the whole exposure is completed within a short time after its injection. (Specifically, since its half-life is about 3½ days, its radioactivity will halve twice in each week, and so will be reduced to about 0.4 per cent of its initial value in only 4 weeks.)

A practical value of these direct risk estimates for bone cancer, is that plutonium, americium, and other transuranic elements that can be significant in occupational and some public exposure, become retained partly in bone (as well as in liver) after being inhaled in soluble form, or – to a lesser extent – if swallowed. The importance of the radium risk estimates is all the greater since they depend on alpha radiation of the bone cells, and are thus immediately relevant to the alpha-emitting transuranic elements.

Thymus gland irradiation

It was at one time supposed that an enlargement of the thymus gland in childhood was responsible for, or contributed to, various forms of difficulty in breathing, or occasionally to otherwise unexplained sudden death. Infants or children who were found by x ray to have unusually large thymus glands were therefore given the moderate doses of radiation, of a few sieverts to the neck tissues, which were found to cause shrinkage of the gland to a smaller size. Although the radiotherapy fields were centred on the thymus, they will ordinarily – and particularly in infants – have included other organs situated in the neck, such as the thyroid and salivary glands.

Very detailed investigations have been made into the subsequent histories of children who were so treated at a number of centres in the northern USA, at which such treatment used to be given during the 1950s. No abnormalities of the thymus were found. An increased number of thyroid cancers had, however, developed, and a lesser increase in cancers of the salivary glands, although few deaths have resulted from either form, since both these types of cancer can be effectively treated in most cases.

In these surveys, the numbers of cancers observed has been compared with the numbers expected in the general population at the same ages and sex. This is likely to be a valid comparison, since it is now doubtful whether the thymus glands that were originally irradiated were in fact abnormal in any functional way. With knowledge of the doses delivered, estimates have been obtained for the induction rate of thyroid cancers in childhood, with a risk of about one per cent per sievert for doses of one or more sieverts to the gland. The risk of causing a fatal thyroid cancer,. however, appears to be about one tenth as great, in view of the ease with which the type of cancer which is caused by radiation can be cured. Salivary gland cancers have occurred with an excess frequency about one tenth that for thyroid cancers, and can commonly also be successfully treated. No excess was detected in cancers arising from other structures in the neck.

Ringworm of the scalp

Radiation has been quite widely used to cause a temporary loss of hair from the scalp, so that the fungal infection of ringworm can be effectively treated. In the course of giving the necessary depilating dose of several sieverts to the scalp, smaller doses are delivered to brain tissues, and much smaller ones – in the region of one tenth of a sievert – to tissues in the neck.

Two large groups of people, who had been treated in this way in childhood in New York or in Israel, were found to have developed a small but definite excess of brain tumours during the 20 years or more since irradiation, the excess frequency being of about one case per thousand treated, the average dose to the brain being a little over one sievert in each survey.

It is of particular importance however that, even at the still lower doses that will have been delivered to the structures in the neck, a small but definite excess of salivary tumours was demonstrable in both series, and of thyroid cancers in the larger Israeli study. In both cases the doses that are likely to have been delivered to these structures, as well as to the brain, were estimated by a careful reconstruction of the conditions and areas of irradiation. Measurements were made of the amounts of radiation reaching the various positions, using models – conventionally, even if surprisingly, called 'phantoms' – of the head and neck, the phantoms being made of appropriate materials to simulate the transmission of radiation through human tissues. These methods indicated an average salivary dose of about 0.4 Sv and a thyroid dose of rather less than 0.1 Sv.

Given the validity of these estimates, and that the children did not wriggle unduly during treatment in a way that the phantoms could not do, the findings are of particular importance, in providing direct – even if only approximate – estimates of risk from doses comparable with, or less than, those liable to be received occupationally during a working lifetime or, in the case of the thyroid, with the dose received by the age of 60 from natural sources of radiation. Moreover, the thyroid risks per sievert at these doses are similar to those observed per sievert at the 20 times higher doses that were given in the course of thymus gland irradiation already described. The risk was only identified at these low doses by the follow-up of large numbers – 11 000 – of irradiated patients, with carefully matched and even larger comparison groups, and with identification of a type of cancer which is ordinarily uncommon in adolescence but which is induced more frequently by external radiation than are most other types of cancer.

Radiotherapy of the breast

Several surveys have been made of the frequency with which breast cancer develops after the treatment of various inflammatory and other benign diseases of the breast in women by radiation. In one such survey, 13 breast cancers were found to have developed during periods of up to 25 years after radiotherapy in the 600 patients so treated, whereas only about 6 would have been expected in women of the same ages in the general population. These findings suggested a radiation induction of cancer in just over one per cent of the patients so treated.

It was possible, however, that cancer might occasionally develop, with a low frequency of this order, as a late effect of the initially benign diseases that had been treated. The reality of the radiation induction of the breast cancers, and the rate of its induction, were however established by further comparison between patients treated for the same disease of the breast in one clinic by radiotherapy, and in another city by methods not involving radiation exposure. Comparisons were also made between the subsequent occurrence of breast cancer in these groups of patients, and in similar numbers of their sisters of an equal total age distribution, so that allowance could be made for any differences in frequency of breast cancer in the two cities, or in families of different genetic origin.

Pelvic irradiation for benign conditions

Irradiation of pelvic organs in women, either by external radiation or by radium implanted temporarily into the pelvis, was formerly used to treat certain benign diseases of these organs, or to bring on the menopause prematurely when this was medically necessary. Follow-up of patients so treated has, in some instances, shown a small but significant subsequent increase in the number of cancers of the intestines, and of other organs partially or wholly located within the areas irradiated.

Evidence following radiotherapy for cancer

For many types of cancer, radiotherapy is usually given at ages at which the normal expectation of life is less than the average time, of 20 years or more, at which radiation induced cancers are likely to develop; and any recurrence of the original cancer in treated patients will further reduce the number of people on whom the necessary prolonged follow-up can be made to detect any radiation induction of cancer. Moreover, if such studies are made, it may be uncertain whether a cancer which develops late after radiotherapy is a recurrence of the original growth or a new and separate cancer which may have been induced by the radiation; and even one or two such uncertain cases could cast doubt on the results of the survey, and on any radiation risk estimate derived from it.

In one instance it seems likely that further estimates of cancer induction at low dose may be obtainable, on the basis of a large international survey of patients treated many years ago for cancer of the cervix of the uterus by radium. In a review of the subsequent medical history of many thousands of such patients, in the clinics in which they have been under surveillance following the initial treatment, slight excesses of various cancers have been detected in organs remote from the position in which the radium was inserted. In these organs the dose of gamma radiation from the radium is small but calculable from knowledge of the amount of radium used, the length of time during which it was implanted, and the distance of the organ from the site of implantation. The very low frequencies of cancer induction resulting from these low doses are only detectable because of the large scale of the survey resulting from the cooperation of many radiotherapy clinics.

EVIDENCE FROM DIAGNOSTIC PROCEDURES

In general, the doses delivered to body tissues from diagnostic radiological procedures are so much smaller than those used for therapy that no increase of cancer has been, or is likely to be, detectable. In three instances, however, such increases have been found, either when many examinations have needed to be made on the same patients, or following exposure of the developing foetus, or – in one case – after the use of a radioactive material as a diagnostic contrast medium.

Multiple fluoroscopic examinations

Active pulmonary tuberculosis used to be treated by the injection of air into the (pleural) space surrounding the lung, so that the affected segment of the lung was collapsed and the spread of the disease controlled. The state of the lungs needed to be examined at about monthly intervals to determine the degree of collapse and the need for further air injection. These examinations were commonly made by fluoroscopy, projecting the x-ray image of the lung onto a fluoroscopic screen so that the collapsed segment could be seen in outline from different directions. The use of x-ray films would have caused lower radiation exposure but offered a less rapid, and in some way less versatile, way of checking the three-dimensional state of the lung segments and the adherence of any of them to the chest wall.

In one clinic, the patients normally faced towards the x-ray tube during these inspections, so that the breast received a higher dose than if they had faced the screen and the doctor. Breast cancer was found to have developed with a significantly higher frequency during the subsequent 15 to 45 year period in women who had needed such examinations than in women at the same ages in the general population, or in women with tuberculosis but who had not required lung collapse therapy. Even though the average radiation dose to the breast was likely to have been only 10 to 20 millisieverts in each examination, the average number of about 100 examinations appears to have involved a large enough total dose for an excess of breast cancers to be detectable statistically, and for a risk estimate for breast tissue to be derived. This estimate is of particular importance for radiation protection purposes. Although the total doses will commonly have exceeded one sievert, the cancer induction should be attributable to the effect of low individual doses of 10 or 20 mSv, since the interval of several weeks between each such dose would have been sufficient for repair of the damage that each produced, had such repair been possible. In the light of present radiobiological knowledge, the risk estimates derived from these data therefore apply to the effects of low doses, in the region of those that are received annually in a few occupations or during 10 years from natural sources. The higher total doses only served to allow the effect of low doses to be detected and measured.

Foetal irradiation

Another instance in which low individual doses appear to have had a detectable carcinogenic effect arose when an extensive survey was reported in 1958, of diseases occurring during the first ten years of life of children whose mothers had had diagnostic x rays of the pelvis during the relevant pregnancies. The radiation doses that these children are likely to have received in this way before birth can only be estimated approximately, from a knowledge of the radiological techniques that were being used at the time, and from the probable number of films taken per examination. It seemed likely, however, that the foetal doses would not have exceeded one tenth of a sievert. Despite this, the number of deaths from malignant disease was significantly increased in these children, the amount of the increase apparently varying somewhat with the stage in the mother's pregnancy at which the examinations had been made.

No such carcinogenic effect of foetal irradiation was, however, detectable in several other studies of children who had been exposed *in utero* to diagnostic x rays, or to radiation from the atomic bombs in Hiroshima and Nagasaki. It was therefore suggested that the observed excess of cancers might be due to some hereditary condition in the mother, for example causing narrowing of the pelvic outlet and requiring pelvic x rays to be taken during pregnancy, with a genetically associated trait in the children causing a small increase of cancer incidence during childhood.

Three pieces of evidence, however, have indicated that the effect is likely to be real, and due to radiation. In the first place, when the findings were re-analysed according to the number of x-ray films that were estimated to have been taken during the examination of the mother, the increased cancer mortality in the children was about proportional to the number of films taken.

Secondly, the cancer rate in twins who had been irradiated *in utero* was raised to about the same extent – both for leukaemia and for all other types of cancer – as it was raised in 'singletons' who had been similarly irradiated. But pelvic x rays were needed, and were taken, much more frequently during twin pregnancies than during others. It can be argued, therefore, that it was the irradiation *in utero* that increased the cancer frequency, and not a factor selectively causing cancer-prone children to be so exposed – at least as regards those factors which involved a high selection of twins for exposure.

A third line of evidence remains equivocal. It was reported in 1962 that the initial results of a large-scale review of the effects of pelvic irradiation in a number of clinics had not only confirmed an increased cancer mortality in children exposed *in utero*. It has also established that, having regard to the size of previous surveys, the various positive or negative assessments of the risk were all statistically consistent with the central value that was found in this much larger and therefore more reliable survey, namely, of a 40 per cent increase in the rather low natural cancer mortality in the first 10 years of life.

The extension of this survey continues to indicate a small increase in the occurrence of leukaemia after foetal irradiation, although the evidence for an increase in other forms of cancer is more doubtful. It may be concluded, therefore, that even at doses lower than about 10 mSv there is some risk, at least of leukaemia, which is detectable by large surveys. The absence of detectable effect in Hiroshima and Nagasaki may perhaps be due to the limited numbers of children on whom the 1970 report was necessarily based.

Effects of Thorotrast injection

The former use of Thorotrast, a preparation of thorium oxide particles, as a radiological contrast medium has already been mentioned, and its effect in causing an increased cancer incidence in tissues in which the particles become deposited and retained. The preparation was used to outline the distribution of arteries into which it was injected, and to detect any local narrowing or obstruction – the thorium having a high atomic number and therefore causing a dark image in x rays taken immediately after its injection.

A number of reports appeared, however, identifying increased rates of cancer, particularly in the two tissues in which insoluble particles typically do become concentrated after they have been injected into the bloodstream, namely the liver and the bone marrow. The corresponding radiation risk estimates for liver cancer and for leukaemia were subject to various uncertainties. Does thorium have a significant carcinogenic effect by a chemical action as well as by its radiation? This now seems unlikely. How much of the weakly penetrating alpha radiation that it emits is absorbed within the particles themselves? This has now been calculated and allowed for. How much of the radiation from a long-retained radionuclide must be discounted in estimating the carcinogenic dose because it is delivered after the cancer has started to develop? This only remains uncertain. The leukaemia risk estimates, however, are consistent with those derived from our sources, and those for liver cancer at least suggest a low sensitivity of the liver to radiation cancer induction.

EVIDENCE FROM HIROSHIMA AND NAGASAKI

A large number of different studies have been made on the medical fate of survivors of the atomic bombing of Hiroshima and Nagasaki.[22]

The most detailed and extensive of these consists in a prolonged and continuing surveillance of the causes of death of about 48 000 people who were exposed to significant amounts of radiation from either of the bombs. This 'Life Span Study' includes also some 35 000 people who were in one of the cities, but who will have had small exposures only, of less than 10 mSv, because of their location at the time of the bombing; and a further 25 000 who came to the cities during the hours or days after the bombing, to help in rescue work or to search for relatives.

The amounts of radiation that were likely to have been received by those exposed to the bombs depended in part on where each individual was at the moment of the explosion, and in part on the amounts of radiation reaching ground level at different distances from the position of the bomb. The first component in the dose estimations, therefore, involved identifying the map position of each individual in the survey, and so the distance from the ground zero below the bomb. It was necessary also to record the type of building in which he was and often the position in the building, to assess the amount of shielding against radiation that the building would afford.

This assessment was necessarily laborious and often approximate, although shielding factors were measured directly by tests with neutron and gamma ray sources on concrete buildings and on houses of the type and materials used in the two cities.

The second component in the estimates, however, required rather detailed knowledge of the proportions of the plutonium or uranium which underwent fission in the weapons, and the resultant quantities of neutrons and gamma radiation released immediately from the fission and from the fission products that were formed. It was necessary also to determine the distribution of energies characterizing these two forms of radiation since this would affect their penetration through the atmosphere, the walls of buildings, and the tissues of the human body, as well as through different parts of the casing of the weapons themselves. In particular, the transmission through air, and so the amounts of radiation reaching different parts of the cities, will have depended upon the moisture content of the air at the time of the bombing.

The importance of these factors has been assessed by a series of tests and calculations, using sources of, necessarily, rather varying degrees of similarity to the exploding bombs themselves. The accuracy of these estimates has recently been improved by a fuller analysis of the amounts and energies of the radiations that will have been released from each type of bomb, and the transmission of these radiations through the bomb casings and through the atmosphere. Overall, these analyses, depending on tests made in the field, observations of radiation behaviour in the laboratory, and detailed calculations of transmission physics, are establishing the conditions of radiation release from the explosions, and the range of doses that will have been received by people at different distances from their detonation. These doses vary from less than 0.1 Sv for those who were over 2 km from the ground zero of the bombs, to values of over 2 Sv for a few thousand members of the group studied, who were at less than 1 km from this point.

The doses received by people who entered the cities after the bombing cannot be estimated with confidence. They are however likely to have been less than 1 Sv even for those entering within a few hours, and less than 0.2 Sv if entry was more than a day later.

Death certificates were examined for almost all of the 20 000 members of the life span study group who had died, and the causes of death were verified in the light of all available clinical or autopsy reports, autopsies having been made in about one fifth of all cases.

This work is of particular importance for radiation risk assessment for a number of reasons. First the large size of the population followed allows small increases in cancer mortality to be detected reliably, and the comprehensive study of medical records makes it possible to identify with adequate confidence the types of cancer which are increased. Secondly, the prolonged follow-up since exposure – already of 37 years in 1982 – ensures that radiation induced cancers or other effects which only develop at a long interval after exposure will be detected. Since the average age of the group that is being studied was about 30 at exposure, estimates are already being obtained of the probability of cancer development during the whole remaining normal life span in many cases. And thirdly, the mortality from different types of cancer can be compared in groups of people exposed at different dose levels. In this way, estimates are not only obtained of the increases in cancer rates in irradiated people by comparison with unirradiated people living in the same city at the same time. Information is also obtained about the way in which the size of these increases varies with the radiation dose received. Estimates have been made in this way of of the risk of fatal cancer induction in some 10 body organs or tissues. These risks vary according to the organ or tissue exposed – from a few cases per thousand following a dose of one sievert in the case of the female breast, the lungs, or the bone marrow (causing leukaemia), to one case or less per thousand after one sievert to stomach, liver, salivary glands, and various other organs.

This study of the frequency with which radiation induces cancers which prove fatal has been supplemented in several ways, to examine the frequency with which non-fatal cancers or other curable conditions are induced. For example, in the 'Adult Health Study' between 15 000 and 20 000 of those who are already included in the life span study of mortality visit a central clinic in one or other of the cities twice a year, and records are maintained of any illnesses, tests, or operations that they undergo. On this basis, a number of reports have been published of the frequency with which various types of cancer have been diagnosed in people exposed at different dose levels. This information has been of particular importance in regard to such cancers as those of the thyroid or salivary glands which can usually be fully removed at operation, and of which the frequency would not therefore be reflected in the mortality statistics of the life span study.

Further evidence of the total induction rate of different types of cancer has been obtained from registers that are maintained of all cases of cancer which have been detected in either city since the bombing. The information from these Tumor and Tissue Registries has the advantage that it relates to considerably larger numbers of people than are included even in the life span study, and may therefore uncover effects that are not detected in that study. It has the weakness, however, that the doses received by the patients in whom diseases are detected have not been estimated individually, as has been done in the life span study. The cancer or other incidence rates are related only to the position in the town where each person was at the time of the bomb, and hence to the average dose at that distance from the ground zero, without the possibility of taking into account the radiation shielding afforded by the type of building in which they were, and by their position in the building.

Despite the variety of radiation effects that can be examined and quantified in these studies, and the large numbers of irradiated people on whom the estimates of risk can be based, it is important to recognize the possible defects of the study as a whole. The estimates of local dose at different distances depend on inferences about the likely release and transmission of different types of radiation from the bombs, aided only by limited measurements made soon after the bombing of the amounts of radioactivity induced in steel, concrete, tiles, and other materials. The doses received by individuals further depend on their position in the open or within buildings, and the location of doors, windows, and upper floors of buildings in relation to the straight line from the exploding bomb. The radiation doses were mainly from gamma radiation, but with a neutron component, now reckoned to be small, which had different characteristics of transmission through buildings and through tissues, and a different effectiveness on reaching the tissues.

It has also been emphasized that those exposed to the bombing represent a selected group, and those who survived the blast and the fires and the high radiation doses in the central areas of the cities are an even more highly selected group. The comparison of cancer frequencies in people who were exposed at high and at low dose, therefore, might not reflect the differences in radiation dose to which they were exposed as much as some late effects of severe stresses, or the characteristics of those who can survive the severer stresses.

It would be easy to imagine such an effect in reducing the early cancer mortality in those exposed at high dose, if people who were already developing cancer were more vulnerable to the immediate effects of the bomb. It is less easy, however, to believe that those who would only develop a recognizable cancer 10 or 20 years later, or who were in some way cancer-prone individuals, would be more easily killed by these effects, and so reduce the cancer frequency in the high dose groups and therefore the risk estimates depending on these frequencies.

It is also possible that unidentified differences in subsequent employment, activities, or lifestyle of those exposed at high or at low dose might determine the occurrence of some cancers. The large variety of different types of cancer which have all been found to be increased in number following higher exposures would seem to make this explanation improbable.

The strength of our present radiation risk estimate for cancer induction in different body organs, however, relies for many of these organs on a broad consistency between the estimates derived from surveys following exposures received under quite different circumstances. Many of these surveys are defective in some respects. Together however they appear to give an approximate but reliable basis for radiation risk estimation. The continuing willingness of so many of the atomic bomb survivors in Hiroshima and Nagasaki to attend the regular medical examinations and surveillance contributes generously to this estimation of the types and the sizes of radiation risks, which is an essential component in all efficient radiation protection.

Seven Marshall Islanders developed thyroid cancers, which were successfully removed by operation, after their exposure to radioiodine in fallout in 1954 (p. 77). For lack of reliable dosimetry, however, no estimates can be made of tissue sensitivity to radiation damage. One member of the crew of the fishing vessel (p. 77) died with jaundice after multiple blood transfusions.

10

RADIATION PROTECTION

Radiation protection depends almost entirely upon preventing people from receiving any undue amounts of radiation, and hardly at all on preventing the biological effects of any radiation that has already been received. Certainly it may be possible to reduce the risk of death or illness resulting from high radiation exposures of the whole body, by transfusions of blood or bone marrow cells and by the prevention of infections. It is also possible to decrease the harmful effects of radiation if certain substances are injected at the time of the radiation exposure or immediately before it, but the usefulness of these 'chemical protectors' is limited by their very short period of effectiveness in the body. Except in a few unusual circumstances, therefore, radiation protection deals with methods of avoiding or limiting exposure, and the verification that exposures are thereby reduced to appropriately low levels.

The situation differs in three main types of exposure:

1. Radiation from natural sources, which causes the greatest total exposure of the population as a whole, cannot effectively be reduced, except in some cases as regards exposure from radon in buildings.
2. In medical procedures, which cause most of the remaining total exposure of populations (p. 91), both the diagnostic and the therapeutic effects depend on body tissues being exposed to the radiation. The objective here must therefore be to obtain the necessary results, while minimizing all tissue exposures which are not necessary for these purposes.
3. In other types of exposure, either of workers or of the public, the processes giving rise to the radiation may be necessary or unavoidable, but the exposure of people to this radiation needs to be minimized.

In view of the rather different approaches to radiation protection that are involved in these three cases, it is useful to review the available methods of dose limitation separately in each.

PROTECTION AND NATURAL SOURCES

The irradiation of all body cells from the normal radioactive component of our body potassium cannot be reduced, since potassium is a constant and necessary body constituent.

The intensity of cosmic radiation varies somewhat with altitude and with latitude, as already described (p. 62). At any given location, however, it would be little diminished even by heavy lead shielding. Similarly, the amount of gamma radiation reaching the body from the soil or from underlying rock varies substantially in different localities (p. 64), but is only moderately influenced indoors by the shielding effect of most housing materials.

On average, however, the largest single component of our exposure from natural sources comes from the radioactive gas radon, released particularly from some types of soil and building materials, and accumulating in the air of buildings in which ventilation rates are low.

In most buildings, the doses resulting from radon inhalation are likely to be less than the total doses received from the other, and unavoidable, sources of natural exposure. Moderately higher than average dose rates, from the radon daughter products deposited in the lungs, are estimated for houses built of granite or other igneous rock. Substantially higher dose rates have been determined, however, in houses built of certain synthetic materials or shales of significant radioactivity (p. 68), or located on land that had been infilled with mine tailings, where radon concentrations have been 50 or more times the national average. Radon levels have been high also in houses with earth floored half-cellars in districts of high soil radioactivity, these levels being moderately reduced by concrete flooring of the cellars. In the same way, the covering of walls with sealant substances may reduce the release of radon from radioactive building materials.

In most countries, the indoor radon concentrations would only occasionally justify remedial measures in existing houses, or the avoidance of new building in particular areas or with particular building materials. Exceptionally, however, such measures may need to be considered, where radon concentrations are likely to be at many times the average value, and at levels capable of causing a significant lung cancer hazard – comparable perhaps with that resulting from smoking a few cigarettes per day.

PROTECTION IN MEDICAL PROCEDURES

It is easy to say that unnecessary radiation exposure in medicine should be avoided. It is often difficult to see in advance, however, how large a treatment dose or how many x-ray films will be needed, and how much exposure will prove to have been unnecessary. Good judgement in ensuring the required result without any undue exposure in one of the skills of the radiologist. A number of measures are important, however, in helping to restrict the irradiation of body tissues to what is essential.[69]

Radiotherapy

In the treatment of cancer, the size and position of the 'field' which is irradiated is matched to the known position of the cancer, but including also any surrounding areas into which it may have spread. Other parts of the body are shielded from irradiation as much as possible. Excessive irradiation of any area of skin may sometimes need to be avoided by directing the x-ray beam towards the tumour through different areas of skin at the time of successive treatments. In addition, the maximum dose to the tumour with the least risk of unacceptable damage to the skin is achieved by using penetrating x rays or gamma radiation of high energy; radiations of lower energy and weaker penetration give up more of their energy to the skin, and the 'depth dose' at the position of the tumour is correspondingly less. A better ratio of depth dose to skin dose can sometimes be obtained by neutron irradiation, since neutrons may give up most of their energy at a few centimetres below the body surface, when they have been slowed down by passage through the surface layers and are 'captured' in the deeper tissues.

Diagnostic radiology[70]

In diagnostic radiology also, it is important to restrict the size of the beam to that part of the body of which the x-ray pictures are needed; and also to shield sensitive structures such as the germinal tissues when they might otherwise receive significant exposures.

Different types of modern x-ray film will give pictures of equally good definition with substantially different exposure of tissues. Methods are also being explored to record and reproduce the x-ray image electronically, without the use of photographic development. Moreover, in CAT scanning, very much more discriminating information is obtained of the position and outline of body structures than by conventional x rays, at the expense of only moderately increased radiation exposure of the tissues through which the scanning x rays pass.

A variety of methods have therefore replaced the old fluorescent screen in obtaining images of body structures at progressively lower exposures. The skill of the radiologist is still essential in judging what types of x ray, what numbers of films, and what angles of view are likely to be necessary; and in limiting the duration and the doses involved in fluorescent screening in those cases in which body structures need to be watched in movement on the screen, rather than simply recorded in position on film.

Nuclear medicine therapy

The objectives of radiation protection in nuclear medicine are similar to those in other forms of radiology, although the methods differ. In the few but important circumstances in which radioactive materials are valuable in treatment, the essential is again to deliver the maximum necessary dose to the diseased tissue, without undue or unnecessary exposure of other body tissues. This is achieved by incorporating an appropriate radionuclide into an appropriate chemical compound. The chemical compound should be such that it is selectively concentrated and retained in the tissue that needs to be irradiated, with minimal or transient retention only in any other important body tissue.

Thus when radioactive iodine is given in the chemical form of iodide, it becomes concentrated – not only in the thyroid gland – but also in some types of thyroid cancer tissue, in which the tissue retains the biochemical behaviour of the thyroid cells from which it has developed. For this reason it is often possible to deliver high doses of radiation to all such cancer tissue present in the body, regardless of the positions to which it has spread, and regardless of whether the sites and extent of this spread are known. The radioiodine that is not so concentrated is rather rapidly excreted from the body, with little irradiation of other body tissues apart from the thyroid gland. The activity of this gland requires in any case to be suppressed to allow treatment of the cancer, and can then readily be replaced by thyroid hormone administration.

In much the same way, radionuclides are injected as colloidal or insoluble chemical compounds into the cavity of joints, to irradiate the inflamed membranes of the joints in certain forms of arthritis. Here the chemical form of the injected material ensures that it remains mainly within the joint cavity, and irradiates only the surfaces of the cavity during its radioactive decay.

It is important also that the radiation emitted by the radionuclides used in therapy shall, as far as possible, have sufficient penetration to reach the diseased tissues that need to be irradiated, but shall not have great enough penetration to cause significant exposure of normal tissues at a distance from the position in which the radionuclide is concentrated. As well as this requirement on the penetration, and therefore the type and energy of the radiation emitted, the radionuclides used should have a half-life of decay that is sufficiently short, so that normal body tissues are not irradiated when the radionuclide is discharged from the diseased tissue in which it was initially concentrated.

Diagnostic nuclear medicine[71, 72]

Two conditions are again needed, to ensure that the required amount of diagnostic information is obtained with the least unnecessary exposure of any body tissue. Firstly, the chemical compound used – the radiopharmaceutical – should be selectively and highly concentrated in the organ of which the position or function needs to be determined; and, secondly, the compound should contain a radionuclide of which the position and the amount can be accurately measured by counters external to the body. The half-life of the radionuclide should be such that adequate radiation is being emitted at the time when the measurements need to be made, without unduly long persistence of the radioactivity, and unnecessary tissue exposure, after the test has been completed. The latitude available in choosing appropriate chemical compounds, and radionuclides of appropriate half-life and penetrating emissions with which to 'label' them, has made it possible to outline the position or measure the activity of a great variety of normal and diseased tissues, with no greater radiation exposures than are involved in conventional x-ray procedures.

PROTECTION AGAINST OCCUPATIONAL EXPOSURES

The need for dose limits

In considering exposure from natural sources, there was no question of deciding upon an administrative limit to the annual amount of exposure that should be regarded as permissible, except perhaps in the case of new construction of houses of materials, or in regions, from which unusually high radon exposures would result. Otherwise it is assumed that living at high altitudes, or on areas of monazite sands or igneous rocks, involves a slight radiation hazard but would no more be prohibited in consequence than would residence in areas at moderately increased risk from earthquake, flooding, typhoon, or volcanic activity.

Similarly in medical exposure, the clinical urgencies of diagnosis or efficient treatment vary so greatly from case to case, and, in the patient's interest, should so much determine the radiological procedures used, that no single limit could properly be set to the exposures that should be received regardless of circumstances.

In reviewing occupational exposures, however, we need to consider not only the ways in which the amounts of exposure can and should be limited in different circumstances, but also the limit that should be imposed on the amount of exposure that should be received by any worker in any year, to ensure that the occupation is of an appropriately high degree of safety in this respect.[30]

It is easy to describe the methods that are used to reduce radiation exposure at work, and the levels to which they reduce it. It is easy to estimate approximately the risk involved in the exposures that are received in this way. It is much less easy to assert what amount of risk should be regarded as acceptable, or what degree of safety should be ensured, in well planned and well regulated occupations. This is no longer a scientific question, but essentially one of societal assessment. It is recognizable as one which arises in any occupation involving exposure to potentially harmful chemical or physical agents, and particularly when it cannot be assumed that there is any safe threshold level of the agent below which no harm is caused. It is, however, or should be, recognized also as a significant question in regard to the risks of every other occupation. The risk of occupational harm, including that of fatal accidents, varies greatly in different conventional industries, but is present and measurable in all of them.

So, how safe are existing industries; and, given that no industry (or other human activity) is entirely safe, how safe must an industry be, to be regarded as reasonably safe?

This is a substantial question, and deserves to be addressed in concrete and numerical terms more often than it has been in the past. It arises also in regard to the proper protection of the public against exposure to radiation and a number of other identifiable harmful agents in the environment. It is worth dealing in the present chapter with the methods available for limiting occupational exposure, and that of the public, to radiation. We can then review the resulting levels of exposure, and levels of risk, in the following and final chapter, with a discussion of the limits of occupational and public exposure that have been recommended as ensuring an appropriately high degree of protection in different circumstances.

Occupational exposure may result either from external radiation of the body from sources used in the industry, or from internal radiation due to radioactive materials entering the body by being inhaled, swallowed, or absorbed through wounds.

External radiation[73, 74]

External radiation exposures are reduced by two methods: by keeping at an adequate distance from the source, and by shielding against its radiations.

The protective effect of distance is obtained if tongs or remotely controlled instruments are used when radioactive materials need to be handled (and 'handled' is therefore the wrong word, and 'manipulated' is not much better). In this way, if the hand is kept at a distance of one metre from the source, rather than at one centimetre if holding it, the dose to the hand is reduced by a factor of 10 000, since the dose received from a source decreases with the square of the distance from it. Similarly when, for example, a radiotherapy source of cobalt-60 is kept within some form of shielding except when in use, remotely controlled methods of withdrawing it from its housing allow the operator to maintain his distance from the unshielded source.

Otherwise, and apart from the obvious uses of distance from any source in the design of working positions and the time spent in them, shielding is of much wider application than distance in occupational protection against external radiation.

Shielding against most of the penetrating forms of radiation, for example from x rays and gamma radiation, is achieved by interposing a barrier of steel, lead, or concrete between the source of the radiation and the positions in which people work. This in itself is not enough, however, since radiation from the source becomes scattered by the air or other material through which it passes, and so can turn corners. Shielding is therefore often required to enclose the source more or less fully, rather than simply to prevent radiation passing on the direct straight line between the source and the worker. Such shielding varies from the massively thick walls of concrete and other materials surrounding a reactor core, to the metal capsules in which industrial radiographic sources are contained when not in use. It was the removal of such gamma-emitting sources from their protective capsules which caused the tragedies in the Mexican, Algerian, and Chinese families described above (p. 99).

For sources of high activity such as those in reactors, reprocessing plants, and departments of radiology, where complete enclosure of the source is impractical or where some amount of radiation passes through even great thicknesses of shielding, the exposure can nevertheless be very considerably reduced in normal working positions in these ways, even if not completely eliminated. However, certain types of work, such as that involving maintenance of reactor equipment or correction of defects, may necessarily entail rather higher exposures than those received in regular operation of the plant.

In some types of occupation, no effective shielding can be installed between the worker and the radioactive source. In uranium mining, for example, the rock walls of the galleries and working faces emit significant amounts of gamma radiation and contribute appreciably to the annual exposure of the miner. The amount of this exposure depends in part on the grade, or uranium concentration, of the ore that is being mined, and on whether mining is from the surface or underground.[75]

In general, however, for most sources of penetrating radiation, a suitable design of shield and of working procedure can ensure that external radiation exposures are reduced to low levels in normal working positions. Efficient shielding against neutrons can also be maintained, for example in reactors or in certain forms of radiotherapy when a high flux of neutrons is generated, although now the shielding materials are different. Neutrons pass readily through lead or steel, but are ‘captured’ by various materials (of low atomic number) such a boron or lithium. Reactor shields therefore incorporate suitable such materials to prevent significant passage of neutrons through them.

Shielding against the more weakly penetrating radiations is relatively simple. Most beta radiation is stopped by a millimetre thickness of lead, whereas some gamma radiation is only halved in intensity by 15 centimetres of lead; and alpha radiation cannot pass through the thickness of a few sheets of paper. With these radiations, therefore, the problem usually is not with external exposure, but with the entry into the body of radionuclides which emit them and so cause internal exposure of body tissues.

Internal exposure

In the use of radionuclides in nuclear medicine, the staff need to take simple precautions to prevent any accidental intake of these materials into the body when patients’ doses are being prepared, or their excretions dealt with. These operations are therefore carried out wearing rubber gloves; and smoking, eating, drinking, and the use of lipstick should not be allowed in rooms in which radionuclides are manipulated and in which benches might have become contaminated.

When liquids need to be measured by being drawn up into pipettes, suction is applied to the pipette by a bulb and not by the mouth. With the relatively low activities of radioactive materials ordinarily used in nuclear medicine, simple precautions of this kind, coupled with measurements to detect contamination of gloves, hands, and working surfaces, can usually prevent significant internal exposures from occurring. When rather higher activities of radioiodine are being used in the treatment of thyroid disease, however, measurements over the thyroid glands of members of the staff can be made to monitor the efficiency of the precautions against accidental intake of radioiodine. The regular use of ordinary iodized salt at home can also reduce the amount of thyroid exposure resulting from any such intake.

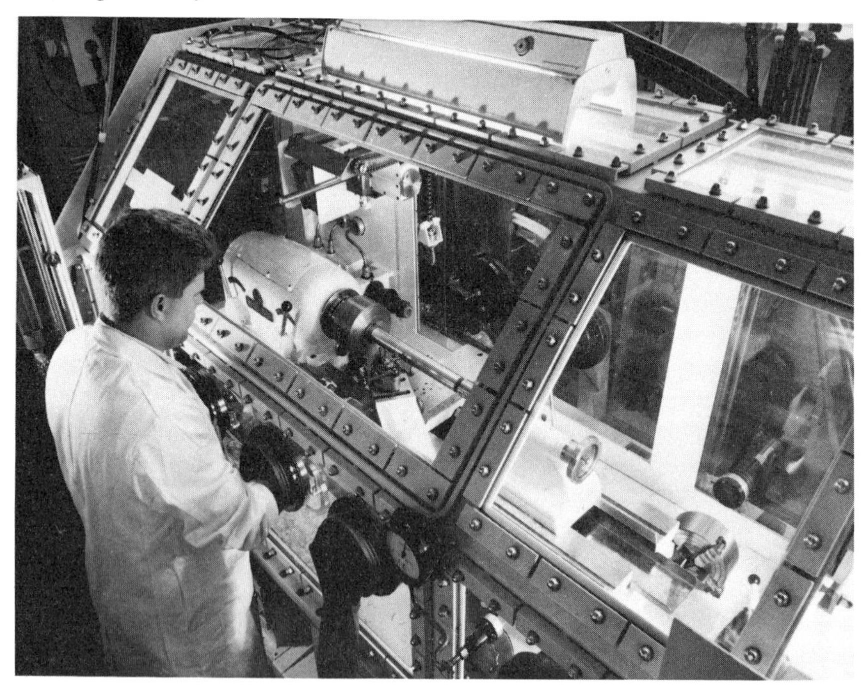

Fig. 10.1. Glove box, for enclosure of radioactive materials during their manipulation. Reproduced by courtesy of Dr Norman Stott and of the Atomic Energy Research Establishment, Harwell.

In many industrial activities, however, which involve the processing of large amounts of radioactive materials, including alpha emitting radionuclides, much stricter precautions are needed. Now the objective must be to keep the radioactive materials fully enclosed during the whole sequence of chemical and other processes to which they are subjected. In this case any manipulations that are needed are carried out by remote control, or with instruments held in gloves which are themselves sealed into the face of the closed 'glove-box' in which the radioactive materials are contained (Fig. 10.1). Attention needs to be given to exhaust ventilation of the glove boxes or other containment and of the working spaces; also to the detection of any defects in the gloves or the containing structure. The air at or near working positions is regularly monitored for radioactivity, as are the workers' hands and clothing when they leave the active areas. It is commonly necessary also to measure the radioactivity of the body or of the excretions, both routinely and at times when any escape of radioactive material into the working environment is known or suspected. The appropriate methods of whole body or lung counting, or of measurements on urine, nasal secretions, or other samples, depend on the sorts of contamination which could have occurred. For example, if the contamination is with water incorporating tritium, the hydrogen-3 radionuclide, of which some is released into the working environment during the operation of heavy water reactors, this radioactive water mixes freely with the stable water content of the body, and so can be reliably assessed by measurement of its concentration in the urine.

Table 10.1

Estimated radon concentrations in the air of uranium mines in the USA

Period	Concentration	Sources of estimate
1937/43	15	Public Health Service
1944/51	11	Public Health Service
1952/58	8	Public Health Service
1959/66	3.5	Public Health Service
1967/69	1.0	Bureau of Mines
1975/77	0.4	Department of the Interior
1979/80	0.2	Department of the Interior

Concentrations are quoted in 'working levels', where one working level corresponds to a concentration of radon-222 in air of 3.7 becquerels per litre (measured in equilibrium with its short-lived daughter products).

For some radionuclides, methods can be used to accelerate the discharge from the body of any intake that has occurred, either by increasing its excretion in the urine, reducing its absorbtion from the gut, blocking the uptake into some body tissues, or releasing it (by chelates) from its chemical binding in body fluids or tissues in which it is already incorporated. Removal of radioactive material from wounds, and even from the lungs, is also practicable.

In uranium mines, and in some other forms of hard rock mining, radon is released from the rock and from the mined ore in the same way as it is released into the air of houses from some soils and building materials. Intake of this radioactive gas into miners' lungs can be reduced, as in houses, by efficient ventilation of the galleries of the mine in which radon would otherwise accumulate. The amount of radon inhaled, and the doses to lung tissues from its radioactive daughter products, can be estimated by using devices which measure the concentrations of radon in the air at working positions. With increasing awareness of the radiation hazard, and with attention to efficient ventilation, the air concentration of radon in uranium mines in the USA has fallen very considerably during the last 40 years (Table 10.1), from levels which caused a readily detectable increase in lung cancer mortality, to values in modern mines which conform to relatively high standards of radiation protection.[76]

The intake of radium into the bodies, and into the bones, of early workers in the luminizing industries has been referred to already (p. 130) as a source of evidence of the production of bone cancer by radiation. The disuse of radium as a luminizing agent has removed this particular source of occupational hazard, although contaminations with the more modern agents still require careful control.

PROTECTION OF MEMBERS OF THE PUBLIC

Apart from their exposure in the course of medical radiology, and that from radon in houses and from other natural sources, members of the public are exposed to radiation in various ways which on average contribute only a small fraction of their total exposure, but which may locally expose smaller groups of people at higher levels. Different protection methods apply against fallout from atmospheric weapon tests, against releases from nuclear installations, and against exposures from the use of various consumer products and other practices.

Fallout

Fallout from atmospheric weapon tests, although world-wide, is now exposing individuals to doses which are judged to be too low to call for the difficult measures which would be required to reduce these doses still further. At periods of the much heavier fallout in the 1960s plans were made and limits were fixed for withholding the use of fresh milk temporarily, if obtained from cows grazing in the open, since appreciable amounts of radioactive iodine were deposited on pastures for a few weeks after each large atmospheric test. The relevant radioisotopes of iodine had short half-lives, however, and concentrations in fresh milk never reached levels at which the appreciable risks involved in abrupt changes in infants' diets seemed likely to be outweighed by the small risks from slight radiation exposure of their thyroids.

Discharges from nuclear installations

Discharges of radioactive materials into the general environment may occur at various stages of the nuclear fuel cycle.

Discharges from reactors

In the reactor, by far the greatest part of the fission products formed by combustion of the nuclear fuel, and the remaining fuel itself, are retained within the fuel rod. Small amounts of these materials, however, escape through defective or corroded areas of the metal cladding which surrounds each rod, and are released into the cooling material – ordinarily gas or water – which circulates round the reactor core to convey the heat formed in the reactor to the turbines which generate the electricity. Also, the high flux of neutrons within the reactor induces radioactivity in the structural components of the core or of the coolant circuit. Some of these materials enter the coolant flow and so pass out of the reactor vessel with it. All reactors have treatment systems to remove any radionuclides which escape from the reactor into the coolant in these ways. A fraction of the radioactive gaseous and particulate materials, however, pass into the atmosphere through the reactor stack; some radionuclides are discharged also in the stream of cooling water issuing from reactors.

Regular measurements are made to determine the rates at which the various radionuclides are being discharged into the atmosphere, or in aqueous discharges to sea or river. The discharge of each type or group of radionuclides is controlled so as to be below the limits laid down by national authorities. These limits are set so as to ensure that the maximum doses that might be received by any members of the public from all components of such routine discharges should be as low as reasonably achievable in the reactor operation; and should in any case not exceed – and preferably should not approach – some stated fraction of the recommended dose limits for members of the public, which are discussed below (p. 178). The elaboration and efficiency of the methods which minimize the escape of radioactive materials from the reactor need to be such that each reactor operates well within these limits of permitted discharges.

Accidental releases from reactors

Much greater releases from reactors could occur if the flow of coolant failed, or became insufficient to remove the heat being generated within the reactor core. In this circumstance, if a significant fraction of the metal cladding of the fuel rods melted, substantial amounts of radioactive material could be discharged into the reactor building and in some cases into the environment.

In estimates that have been made of the amount of biological harm that could result from meltdown of different proportions of a reactor core, it has commonly been assumed that a major part of this harm would result from releases to the environment of radioactive iodine in volatile form. It is of interest and some reassurance in this respect that, although a substantial amount of radioactive iodine was released from the reactor core during the accident at Three Mile Island in 1979, almost all of it became 'plated out' within the reactor building itself and did not escape to the general environment, although gaseous radioactive materials were so released.[51] If this type of retention of radioactive iodine is likely to occur also following melt-down in other circumstances, it would reduce considerably the actual number of harmful effects compared with the number that have been estimated as likely to follow some forms of reactor accident. It would also diminish the importance of supplying stable iodine tablets to suppress the thyroid uptake of radioiodine in the event of such releases.

Meanwhile, however, the use of these tablets could prevent much of the thyroid gland irradiation that would otherwise result from inhaling radioactive forms of iodine during any accidents which involved the atmospheric releases of these radionuclides. The tablets contain enough of the normal, stable form of iodine (typically about 0.1 gram) to satisfy immediately the thyroid gland's daily need for iodine. As a result, the thyroid takes up a much smaller proportion of any radioactive iodine that has been taken into the body, and the irradiation of the thyroid is correspondingly reduced. What is more, the uptake of radioiodine into the gland is arrested within minutes of the stable iodine tablet being swallowed.[77] Even if the tablet is only taken 4 or 5 hours after radioiodine had been inhaled, the thyroid receives only half the dose of radiation that it would otherwise have received.

Other protective measures available in the event of any major accidental release vary from the simple precaution of remaining indoors with windows shut during a period when airborne radioactive material is still being discharged and is passing by, to possible later restrictions on the use of fresh milk from pasture-fed animals or, in extreme cases, to a restriction on the use of land for production of crops or other produce or even on occupancy of any housing or land which was heavily contaminated with long-lived radionuclides.

Reprocessing plants

Although small amounts of radioactive material escape from the fuel rods while they are in the reactor, most of these materials still remain within the spent fuel rods when they are removed from the reactor for reprocessing or disposal. Further leakage may occur, however, during the period of some months during which the rods are normally stored, in tanks through which water is circulated, in order to allow the shorter-lived fission products to decay before reprocessing of the unused fuel. Corrosion of a number of spent fuel rods from Magnox reactors developed at this stage while the rods were awaiting reprocessing at Windscale, with some escape of fission products into the water circulating through the tanks.

This release resulted in a substantial escape of two radioisotopes of caesium into the sea during a period of several years[78] before a satisfactory effluent treatment plant was established to remove a large proportion of them from the water before its discharge to the sea. During this period detectable human contamination with radiocaesium resulted in people who were eating locally caught fish; the amount of their radiation exposure is described below (p. 178).

Some discharges, however, occur regularly during the processes involved in opening the spent fuel rods, and treating their contents chemically to separate plutonium and unused uranium from fission products and other waste material. These processes are carried out in shielded and enclosed systems, and chemical methods are used to remove as much of the radioactive material as practicable from any gaseous and liquid discharges. For some radionuclides, however, such as those of the chemically inert gas krypton, no adequate or efficient methods are currently available, and different amounts of various radionuclides are finally released to the atmosphere or to the sea or rivers. These amounts are regularly measured or estimated and, as with discharges to the environment from reactors, require to be kept within limits determined by national authorities.

Waste disposal

Nuclear wastes vary in activity from surgical dressings, protective clothing and other materials which are lightly contaminated with radionuclides of short half-life, to highly radioactive materials including alpha emitting radionuclides of very long half-life. The precautions needed to ensure that their storage or disposal does not result in any substantial human exposures, either as individual or collective doses, vary accordingly. For all levels of activity, sound radiation protection needs to be achieved by a suitable combination of methods of storage or disposal: storage, whereby wastes are kept under controlled conditions which prevent significant escape to the environment; or disposal, when wastes can be discharged in such a way that any return to the general environment is prevented or delayed until the remaining radioactivity of the wastes is low. It seems clear that, in general, if such methods ensure good protection of human individuals, they will also ensure adequate protection, and at least maintenance of numbers, of populations of other species, even though doses and the probability of harmful effects will sometime be higher in individuals of those other species.

With wastes of low activity, storage is rarely needed unless long-lived alpha emitting radionuclides are present, and conventional methods of industrial or domestic waste disposal can ordinarily be shown to involve no likelihood of any significant human exposure. The aqueous and atmospheric discharges of the small unretained fraction of radionuclides arising in reactors and during reprocessing normally cause only limited exposures, either locally or collectively (p. 83). The rather greater exposures due to the release of radioactive caesium from plant at Windscale resulted from delay in reprocessing accumulated fuel rods and in installing appropriate effluent treatment plant, rather than because of any technical difficulties in preventing this type of exposure.

Similarly the disposal of solid waste, contaminated at low or moderate activity, ordinarily requires only conventional industrial procedures, with or without preliminary storage, but often with methods to reduce the bulk of the waste. According to their activity and content, these wastes may be incorporated into materials such as concrete, bitumen, or plastic resins to minimize their release from their containers, or from sites of disposal into excavated trenches, deep mines, or ocean floor. Estimates of the collective doses resulting from the controlled disposal of these low or moderate activity solid wastes indicate that these doses make a minor contribution only to population exposure from nuclear activities.

The disposal of wastes of high activity containing long-lived alpha-emitting radionuclides, requires that a number of barriers should be placed between the radioactive materials and the pathways by which they might ultimately return to the immediate human environment.

These barriers should prevent or delay the escape of these materials

(1) from the glass, synthetic rock, or other solid form in which they are incorporated, due to leaching of their surface layers;

(2) from within the containers in which these solids are placed, as a result of corrosion of their metal structure;

(3) from the cavities in the rock, salt or other formations in which the containers are deposited, by water flow through these formations; and

(4) through the layers of earth or ocean lying between the original site of deposition and any significant source of human food, water, or inhaled air.

The human exposures that might ultimately be received depend upon how long and how completely these barriers are effective. Within 200 years these wastes will have lost most, typically about 99 per cent, of their initial radioactivity, following the decay of most of the fission products which they originally contained. Their radioactivity would still, however, be some 50 to 100 times greater than that of the uranium ore from which they had been derived, and their subsequent loss of activity is slow.

Protection against exposure from these wastes depends, therefore, on incorporating them into a solid material in which they would be retained as stably as uranium is retained in rock; and, as an additional precaution, placing these solid blocks, in water-resistant containers, into formations through which there is too little flow of water to return to the surface any significant amounts of long-lived nuclides which become leached from the blocks and escape from the containers. The former objective may be achievable by fusion of high level waste, either into heat- or water-resistant types of boro-silicate glass,[79] or into synthetic forms of rock of which the constituent minerals are of types known to remain stable through long periods of geological time.[80] In each case, the wastes would not be incorporated into the solid form for 50 years or more, until their radioactivity, and therefore their generation of heat, had fallen to a level which was known not to affect the stability of the material into which they were fused, or the containers into which this material was placed.

Measurements have been made on the rate at which layers of metal used to encase these containers would be dissolved away by water flowing over them, and with some metals these rates are very slow. The thickness of a layer of titanium, for example, is found to be reduced only by 0.0013 mm per year by a continuous flow over it of saline water such as deep ground water would be.[81]

On this basis, a half inch thickness of this metal covering the container should itself last for a period of the order of ten thousand years. A further suggested thickness of four inches of lead should, in the case of waste from reprocessed fuel, ensure that the steel of the container would not be exposed, even if water had flowed past it continuously, until both the radioactivity and the radiation toxicity of the contents were similar to, or below, those of the uranium ore from which the wastes had been derived. (At this stage, the wastes from unreprocessed fuel, with their higher content of alpha emitting radionuclides, would be of about 10 times greater toxicity.)

The waste containers would need to be placed into cavities in a rock or other formation through which water flow was minimal – both to delay corrosion of the metal containers, and, more importantly, to slow or prevent transport to the surface of radioactive material which might escape from this containment. Geological evidence on the date at which various salt formations were laid down shows that no significant amounts of water can have flowed through them, at least during the past millions of years, so that no removal from such sites would be expected. For deposition of containers either in crystalline rock formations or in salt domes, the deposits should be made at some hundreds of metres depth, both to minimize the likelihood of disturbance by natural erosion or human agencies during ensuing centuries, and to decrease further the chances of any significant return to surface waters. The passage to the surface in itself is likely to impose a considerable barrier, since many chemical substances are strongly adsorbed onto clay or other soil elements. The rate of movement from sites of deposition in crystalline rock of various alpha-emitting elements (including plutonium, uranium, neptunium, and radium) has been studied and found to be substantially less than one thousandth that of the flow of water in which they were present; and direct evidence for the immobility of plutonium over very many millions of years has been referred to already (p. 43).

Other stages of the nuclear fuel cycle

Protection measures will, in the future, need to be taken in regard to two other stages of the nuclear fuel cycle: one concerned with the tailings from uranium mines and mills, and the other with the dismantling of nuclear plant at the end of its working life.

The tailings which accumulate during the mining of uranium and the milling of the ore contain little remaining uranium, but most of the radium which is formed by its decay. This radium-226 has the long half-life of 1600 years, and forms radon-222 by its own decay. Release of radon from the tailings pile would be liable to cause exposures at a low rate but continuing over very long periods of time.

Radon-222, however, has a rather short half-life of only 3.8 days. The amount of radon escaping from the tailings could therefore be reduced by covering the tailings pile with a layer of earth or other material sufficient to delay the diffusion of this gas to the surface by long enough for most of the radon to have decayed to non-gaseous daughter products. It has been shown that the amount of radon issuing from the surface would be reduced by a factor of 4 or 5 by every metre of soil used to cover the tailings,[82] or by more when the soil was moist.[83] A covering of about 6 metres of soil would therefore be likely to reduce the radon emission by a factor of 200, and so bring it down to that occurring naturally from the soil in the absence of mining or tailings. Some such stable covering of the tailings from worked-out mines, with reliable sealing of these areas to prevent any significant migration of radium into local water tables, or backfilling into the mines, should prevent any increase of exposure of near-by population above that received normally in the district.

Experience has been gained in a number of countries on the dismantling of small reactors and of two reprocessing plants. This experience suggests that no undue exposure of workers or of the public should occur during the decommissioning of the larger present power reactors, provided that the various radioactive components of the structure are removed successively over a period of some tens of years. Dismantling would require remotely controlled heavy equipment, and chemical decontamination with control of the resultant wastes. Access to the site would need to be restricted during the whole period of these processes. The mechanical problems of dismantling could probably be made somewhat simpler, and the potential doses to workers thereby reduced, by full consideration of these problems during the design of new reactors.

Other sources of public exposures

Most other components of public exposure to radiation are individually small, and small in aggregate. They need to be reviewed, however, in regard to the usefulness of the relevant sources and the extent to which the small risks from exposure to them can be justified, or could readily be reduced. Thus, luminous watches delivered small but significant doses to large numbers of people at a time when radium was the only available luminizing agent. The use of materials emitting less penetrating radiations has made the exposures even lower, without decreasing the luminosity of the watches. This development can thus be regarded as a good radiation protection measure, even though the use of radium was justified, in risk-benefit terms, when no other agent was available.

On the other hand, x-ray machines used to be common in shoe shops which allowed parents and staff to view the image of the children's feet as their new shoes were being tried on, and allowed the children to enjoy the sight of their wiggling toes for as long as they could get away with it. Here the very dubious benefit or necessity of the x ray in shoe fitting was clearly outweighed by the very definite risk of impaired bone growth following repeated irradiation. A close scrutiny and control is called for, and is now undertaken in many countries, of the risk-benefit balance and justification for a whole range of consumer products containing radioactive materials, from smoke detectors and alarm clocks to luminized fish floats, uranium weighted golf balls, and dollies with luminous eyes.

The cosmic ray exposures received during flight are usually small and, as already mentioned (p. 89), are ordinarily about equal for the same distance flown at supersonic as at subsonic altitudes. Very occasionally however, during the development of major solar flares, the dose rate at high altitudes rises rapidly. As a radiation protection measure, therefore, supersonic commercial aircraft are now equipped with radiation rate meters so that, if such a flare developed, the flight could if necessary be diverted to a lower altitude at which the dose rate would be considerably reduced.

One other source of public exposure deserves mention, if only because it has so commonly been overlooked. It has long been recognized that long-lived radioactive materials, originally contained in coal, are present in the fly ash discharged from coal fired power stations. The radiation doses to people from these materials are small, although reckoned per unit of power output from the station, probably about half those from the atmospheric discharges from nuclear reactors (with a regional collective dose annually of perhaps 2 man-sieverts per gigawatt year from many coal fired plants and of about 4 man-Sv from nuclear ones).[23] The low value is due partly to the efficient retention of most of the fly ash formed in modern coal fired stations, and partly to the low population density ordinarily present round such stations.

It is now recognized, however, that larger total collective doses may result from discharges of fly ash from the domestic use of coal. Much less coal is ordinarily used in this way than in power plants, but a considerably greater proportion of the fly ash is discharged, and discharged locally from low chimneys into areas of higher population density.[23] The establishment of 'smokeless zones' may carry benefits even to posterity.

11

SAFETY AND RISK

How safe are the occupations in which the work involves exposure to radiation? Do the precautions and practices discussed in the last chapter give adequate protection against harm, or are there serious risks in present levels of occupational exposure, as there certainly were to the pitchblende miners of the last century and to the early radiologists at the beginning of this one? And how much do the man-made sources of radiation in the environment add to the risks of life?

RADIATION RISK ESTIMATION

For the kinds of non-stochastic damage discussed in Chapter 7 and attributable largely to cell killing, the risks of significant harm are zero unless certain threshold doses are exceeded. Except under accident conditions, no worker is likely to receive exposures from external radiation which approach these thresholds in any organ. Internal radiation is also most unlikely to cause any threshold doses to be exceeded, unless perhaps following a substantial accidental intake of a radionuclide which has a long retention time in the body. The risks of radiation to the worker or to his descendants, therefore, result essentially from the possible causation of cancer or of genetic defects for which no threshold can be assumed, and of which the probability depends on the annual doses received.

Reviews of the somatic and genetic risks

The magnitude of the risk of causing different types of cancer or genetic defect by different amounts of radiation exposure has been reviewed and assessed by a number of national and international scientific bodies over the last 25 years, as increasing information on these risks has been obtained. The British Medical Research Council examined the current evidence first in 1956,[84, 85] and scientific groups convened in the USA reported in detail on this subject in the same year and subsequently.[86] The last of these studies, by the BEIR committee (on the Biological Effects of Ionizing Radiation) in 1980, reviewed very fully the frequency of these effects and the doses causing them. The most extensive reports, however, updating the risk estimates and their implications for different aspects of radiation protection, have come from UNSCEAR and ICRP, the two international bodies already referred to repeatedly (pp. 15, 58, and 61).

Each consists of a group of scientists from a wide range of expertises in radiobiology, medicine, genetics, epidemiology, and radiation physics. The

United Nations committee meets annually with its scientific secretariat – usually with 70 or 80 scientists as members of its 20 national delegations – to develop reports to the General Assembly on the current levels of radiation exposure from all sources, and on the frequency and types of harmful effect which may result from such exposures.

The International Commission and its four committees (dealing with radiobiology and genetics, internal dose estimation, medical exposures, and the application of protection procedures) have a similar total membership, of 76 experts in different fields of radiation, currently drawn from 17 countries. The Commission and the committees call also on task groups appointed ad hoc to examine special problems requiring detailed review and report.

The publications of the Commission,[30, 32] and those of UNSCEAR in 1977 and 1982,[29, 23] give rather exhaustive analyses, of considerable scope and authority, of the present evidence on radiation exposures and effects, and on procedures for effective radiation protection. The International Agencies such as the World Health Organization, the International Labour Organization, the International Atomic Energy Agency, and the United Nations Environmental Programme are normally represented at meetings of both bodies, and use their reports and recommendations in developing the radiation protection criteria and procedures that are relevant to their own fields of work.

Inference to risk estimates at low doses

The risk estimates derived from surveys of irradiated groups of people have been based mainly on the results of exposure of the whole body or of individual organs to doses of one sievert or more, or of about 0.1 Sv in a few cases, as described in Chapter 8. Except in the course of radiotherapy, however, people ordinarily receive much less than one sievert, even during a whole lifetime. Thus the average annual exposure of the UK population is currently of about 3 millisieverts from all sources, or about 0.2 sieverts in a lifetime. The average for all occupational exposures is typically of about 4 mSv per year, delivering a further 0.2 Sv in a working life. Diagnostic x rays and nuclear medicine tests ordinarily involve doses of 1 mSv or less per investigation (as the equivalent 'effective' dose to the whole body, p. 48). It is therefore necessary to decide whether the risks per millisievert are likely to be one thousandth of those per sievert, or more, or less.

The answer is clearly different for high LET exposures, such as those from alpha and neutron radiation, as compared with those from low LET exposures, from x rays and from beta and gamma radiation.[87] For the former, the risk is likely to be simply proportional to the size of the dose. A one per cent risk of fatal cancer from exposure of the body to one sievert would probably therefore correspond to a 1 in 100 000 risk from one millisievert, although with certain forms of high LET radiation exposure, experimental evidence suggests that this simple proportionality may somewhat underestimate the risk at low doses, perhaps by a factor of about 2.

For the commoner forms of exposure, to low LET radiation, the evidence continues to indicate that a risk of one per cent from one sievert would imply a risk of rather less than 1 in 100 000 from one millisievert, and perhaps quite considerably less.

The risks of very low doses of a few millisieverts are unlikely ever to be measurable, or measurable with any accuracy, by direct comparisons of human populations irradiated at doses differing by these small amounts. This, as has now become evident, is because cancers induced by radiation cannot be distinguished by their behaviour or microscopic appearance from other, spontaneous, cancers of the same types. Their frequency can therefore only be distinguished statistically from the frequency that would have occurred naturally. While it has been possible, in a variety of surveys, to estimate the number of cancers caused by doses of one sievert to the body or to an organ, to do so for doses of one thousandth of a sievert would, on statistical grounds, require study of an irradiated population not a thousand times, but a million times as large.

There are, however, two ways in which the risk from very low doses can be inferred, with reasonable reliability, from the risk actually observed at higher doses. One depends upon the way in which the frequency of induced cancers decreases as the dose decreases, within the range of doses over which the excess number of cancers is large enough to be detected. The other depends on our knowledge of the way in which cancers, or inherited abnormalities, are likely to be caused by radiation.

The first method depends upon the shape of the graphs expressing the relationship between size of dose and frequency of effects. For the high LET radiations it is ordinarily found that this 'dose-effect relationship' is represented by a straight line directed towards the 'origin' (of zero effect at zero dose), indicating that the frequency of effects remains about proportional to the size of the dose as the dose becomes progressively lower (Fig. 11.1). For low LET radiations, however, the graph of this dose-effect relationship is more commonly found to be curved, and concave upward, indicating that for these radiations the number of effects that are caused per unit dose decreases as the size of the dose decreases towards zero.

The form of this curve therefore suggests that the frequency of effects at one millisievert will be less, for these radiations, than one thousandth of the frequency observed at one sievert, but it does not by itself indicate by how much it would be less. It is however, clear from a number of experimental observations that, as the size of dose is reduced, low LET radiation becomes considerably less effective – often by a factor of 5 or more – than high LET radiation per unit of dose. Since both forms of radiation are taken to be equally effective per sievert at higher dose (p. 46), and if high LET radiation remains about equally effective per sievert at low dose, it follows that low LET radiation must be rather less effective per sievert at low than at higher dose.

This is to be expected if radiation causes harm predominantly by damage to one or both strands of the DNA helix. If, for example, correct repair is unlikely when both strands are broken at the same moment, the number of incorrectly repaired cells would be proportional to the dose in the case of high LET radiation, since the ionizations caused by these radiations are closely spaced along their tracks, and a single track would be likely to score hits on both DNA strands. With the number of double hits proportional to the number of tracks, the frequency of harmful effects would be simply proportional to the size of the dose.

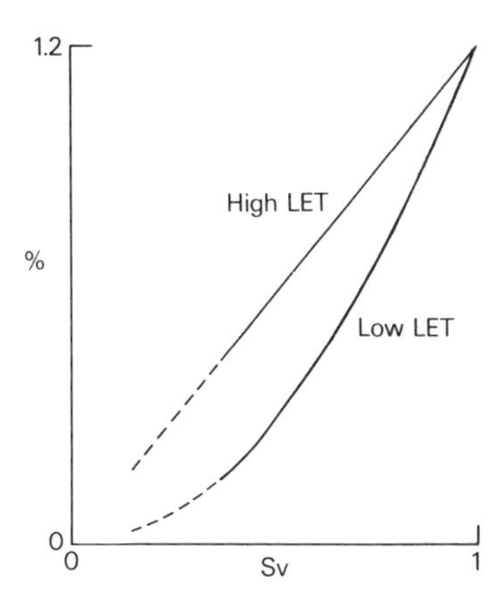

Fig. 11.1. Diagram illustrating frequency of effect linearly proportional to dose for a high LET radiation, and proportional to the square of the dose for a low LET radiation.

With low LET radiation, however, with ionizations much more widely spaced along the tracks, adjacent hits on both DNA strands would usually have to depend upon a second track happening to pass through the same region of the cell nucleus soon after the first track, damaging the second DNA strand before the damaged first strand had had time to be repaired. The number of incorrectly repaired cells would therefore depend on the coincidence, in time and space, of two separate tracks. The frequency of harm would then be proportional to the square of the dose, and the harm per unit dose would decrease progressively as the dose decreased (Fig. 11.1).

Our knowledge is very incomplete on the exact mechanisms of radiation damage to DNA, and on the way in which such damage, if incorrectly repaired, may finally be expressed in cancers or inherited abnormalities. At present, however, it seems probable that for alpha or neutron radiation, the assumption of a simple proportionality between dose and risk of effect gives a good estimate of the risk at low doses, although in some circumstances it may under-estimate it by a factor of about two. For most other forms of radiation, the assumption of a simple proportionality may well somewhat over-estimate, perhaps by a factor of 2 or 3 or possibly more, but is unlikely to under-estimate the risk at low doses. There appear to be no reliable studies on human populations, or on the genetic effects of radiation, which conflict with this probability.

To the accuracy necessary for forming a general assessment of the safety or hazard of different forms of radiation exposure, which are most commonly from low LET radiation, the assumption that the risks of harm are proportional to the doses received is likely to give an adequate estimate of the harm incurred in most circumstances, although this approximation may well over-estimate the harm involved when large numbers of people are exposed to very small doses.

Table 11.1

Effects of exposure of one million people of all ages to 1 mSv each of whole body radiation (assuming simple proportionality between dose and frequency of effects)

Organ exposed	No. of genetic effects		
Gonads	6		(all generations)
	No. of cancers		
	fatal	curable	
Bone marrow	2	0.1	(leukaemia)
Lung	2	0.1	
Breast	5	3	(in women; none in men)
Thyroid	0.5	10	
Bone	0.5	0.2	
Other organs	5	2	

Total

Genetic effects	6 (av. value: fewer for women than for men)
Fatal cancers	15 in women, 10 in men
Curable cancers	15 in women, 12 in men

On this basis, Table 11.1 gives the estimates that have been reached by surveys of the available evidence on the frequency with which genetic effects or different forms of cancer are induced by radiation. The table indicates the estimated number of such effects that might result from the exposure of a million men or women to a dose to the whole body of one millisievert. The cancer frequencies differ somewhat in men and women owing to the induction of breast cancers. There may also be a somewhat greater number of thyroid cancers induced in women than in men, although these cancers, of the types induced by radiation, are usually curable. On the other hand, two major forms of genetic effect (dominant mutations and unbalanced chromosomal translocations) are several times more likely to be induced in mice by irradiation of males than of females. If a similar proportion applies in human genetics, the total number of serious effects of radiation of men and of women would therefore be about equal.

Some skin cancers are also likely to be caused, making a small addition of perhaps 0.1 or 0.2 to the total of fatal cancers, but the skin tumours that are caused by radiation are of types which are so easily cured that no accurate records of their numbers in irradiated populations have been obtained.

The risk of cancer in irradiated individuals and, obviously, that of transmitting genetic abnormalities to later generations, depends upon their age at the time of exposure. The risk of causing genetic effects would fall from about twice the average values given in Table 11.1 for exposures at ages less than 16, to almost zero at ages above about 45. The cancer risk also varies somewhat with age at exposure, and the values given are averages for all ages as observed in the surveys of irradiated populations. Insofar as the frequency of effects is truly proportional to the size of the dose, however, the values given for the number of effects caused by exposure of a million people to one millisievert represent also the number caused by any collective dose of 1000 man- (or woman-) sieverts.

SAFETY OR RISK OF OCCUPATIONAL EXPOSURE TO RADIATION

Most countries adopt the recommendations of the ICRP[30] that all exposures shall be kept as low as is reasonably achievable and that no individual should receive an occupational exposure of more than 50 mSv in any year. The combined effect of these two restrictions is that the average annual dose is usually held considerably below this annual 'ceiling' of 50 mSv, just as the design of houses ensures that we live below the ceiling and not on it. In fact, in countries in which assessments have been made, the dose received, as averaged over all forms of occupational exposure, usually lies between 3 and 5 mSv per year. In these circumstances, therefore, the average dose received at work is about twice that normally received from natural sources. A year's exposure of 4 mSv might then, on the basis of the values given in Table 11.1, involve a risk of causing 4 deaths from cancer in every 100 000 men so exposed, or 6 such deaths in 100 000 women. On this basis 50 years of occupational exposure at this dose rate would raise the percentage of men who die from cancer from its present normal rate in the UK of 23.6 to 23.8. In women the normal percentage of 20.8 would be increased to 21.1.

In some occupations, however, or sections of occupations, average annual rates are higher, and are sometimes in the region of 30 to 40 mSv per year. In the predominantly male working populations in industrial radiography or uranium mining, in which such average rates have been recorded, the total dose received during a 50 year working life might raise the likelihood of dying from cancer from 23.6 per cent to between 25 and 26 per cent. How do these levels of risk compare with those in occupations which do not involve exposure to radiation?

At first sight, the obvious way to obtain some perspective on this question is to compare like with like, and see what increases in cancer rate have occurred in other occupations. The difficulty here, however, is that occupational cancer rates that are only slightly increased would be as difficult to distinguish from the normal rate as they are if caused by radiation, unless very large working populations are studied over long periods, or the cancers were of an unusual type. In some occupations involving exposure to chemicals, increases in cancer rates have certainly been detected in the past, which are up to 40 times as great as the largest increases that can be estimated for work involving present levels of radiation exposure (Table 11.2).[88] Some are greater even than the forbiddingly high rates suggested by statements that 30 per cent or even 75 per cent of early pitchblende miners died of lung disease, assuming this to have been lung cancer. But it does not help us to find that present day occupational exposures to radiation are considerably safer than chemical or other cancer-producing exposures have been in the past in obviously unsafe occupations.

Table 11.2

Cancer mortalities attributed to occupational exposures to chemical agents

Industry, or sections of industry	Year reported	Types of cancer	Fatal cancers (per 100 000 per year)
Shoe making (press and finishing rooms)	1970	nasal	13
Printing trade	1972	lung and bronchus	20
Work with cutting oils (Arve District)	1971	scrotal	40
Asbestos workers (males, smokers)	1972	lung	75
Rubber mill workers	1965	bladder	340
Mustard gas manufacture (Japan, 1929–45)	1968	bronchus	1040
Nickel workers (pre-1925)	1970	lung and nasal	2200
Beta naphthylamine manufacture	1954	bladder	2400
Estimated rates for radiation exposure			
at 4 mSv per year			5
at 40 mSv per year			50

Based, with permission, on data reviewed in *Community Health*.[88]

Nor can the comparison be made any more easily by death rates from other types of occupational disease, and for much the same reasons. Increases in mortality have been detectable mainly when they were large and were clearly attributable to occupational causes. In such cases the size of the increase normally prevents the occupation from serving as a criterion of appropriate safety.

In most occupations small increases would be difficult to detect, or to be attributable clearly to occupational causes if detected.

Table 11.3

Fatal accident rates (annual deaths at work per 100 000 employed) in industries in the UK (average rates for years 1974-1978 except as stated)

Manufacture of clothing and footwear	0.5
Manufacture of vehicles	1.5
Manufacture of timber, furniture, etc.	4.0
Manufacture of bricks, pottery, etc.	6.5
Chemical and allied industries	8.5
Shipbuilding and marine engineering	10.5
Agriculture (employees)	11
Construction industries	15
Railway staff	18
Coal miners	21
Quarries	29
Non-coal miners	75
Offshore oil and gas (1967-76)	165
Deep sea fishing (accidents at sea only, 1959-68)	280

Reproduced, with permission, from material submitted to a Royal Society study group.[89]

Fatal accidents due to occupational causes are, however, readily identifiable and their frequency is regularly recorded in many countries and for many occupations. Reliable figures can be given of the annual risk of death from this cause,[89, 90] as it varies from the safest, to the less safe, and on up to the obviously hazardous occupations; and the range of variation is indeed very large amongst different conventional types of work (Tables 11.3 and 11.4). In this respect, such mortality rates, when expressed in comparable terms, for example of accidental deaths per year per 100 000 employed and at risk, offer one criterion – and an obviously important one – of the relative safety or risk of different occupations. It is manifestly an incomplete and imperfect criterion: the frequency of non-fatal accidents, of fatal or non-fatal diseases, and of injuries causing permanent partial or total disability are also relevant, as is any anxiety of the worker or the family about the safety of the working conditions. It does, however, give us some guide, in numerical terms, of a major aspect of the safety of the kinds of work which we ordinarily regard as being reasonably, or highly, safe – a degree of safety with which we would hope that work with radiation should conform, and particularly in regard to the mortality that it might cause.

Certainly this cannot be a simple numerical comparison. There will be differences in the 'weight' that different people will attach to deaths occurring in different ways and at different times of life. The accidental deaths will often be immediate, and occurring at any time during working life: the average age of deaths from fatal accidents at work is in the early 40s, with loss of some 30 years from the normal expectation of life.[91] On the other hand, deaths from the induction of a fatal cancer by occupational exposure would occur, again as an average, in the early 60s owing to the long interval between a radiation exposure and the appearance of any cancer that it causes.[91] The loss of life expectancy will therefore average about 10 years, and will occur at an age when commitments to a young family are less; but death will be the result of a distressing and sometimes long and painful illness. These two forms of death cannot be simply equated. It is in fact by no means obvious what our choice would be, if offered the alternative at an age of 40: a quick death tomorrow or an uglier death from fatal disease developing in 20 years time. Our decision might depend on whether the question was a comfortably theoretical one or whether we believed that our selection would really happen.

Table 11.4

Fatal accident rates (annual deaths at work per 100 000 employed) in industry groups in the USA

	1955	1961	1968	1974	1980
Trade	12	9	7	6	6
Manufacturing	12	11	9	8	8
Service and	15	13	12	10	7
Government		13	13	13	11
Transport and public utilities	34	43	38	34	28
Agriculture	55	60	65	54	61
Construction	75	74	74	63	45
Mining and quarrying	104	108	117	71	50

Based, with permission, in part on data reviewed in *Community Health*,[88] and derived from *Accident Facts*.[90] These accident rates show reductions averaging about 2.5 per cent per year.

It is equally difficult to make any realistic balance between an accidental death at work and a genetic defect in one of our descendants, or to take proper account of the non-fatal accidents, cancers, or other diseases occurring in different industries, and we will need to return to these questions later. It remains true, however, that practices involving radiation exposure, in which there is inevitably some risk, however small, need to be scrutinized in the light of other working practices in which also there is always some risk however small, to form some assessment of the relative levels of safety in each. In this situation the ICRP has recommended that the practices and dose limits of occupational exposure to radiation should ensure that 'the calculated rate at which fatal malignancies might be induced . . . should in any case not exceed the occupational fatality rate of industries recognized as having high standards of safety'. These industries are 'considered to be those in which the average annual mortality due to occupational hazards does not exceed 10^{-4}', a limit equivalent to 10 deaths per year in every 100 000 workers. This criterion would ensure a degree of safety corresponding to that of most manufacturing industries in the UK and the USA, and better by a factor of 2 to 10 or more than that in occupations such as mining, construction, or trawling (Table 11.3).

In most occupations involving radiation exposure, this limit is not exceeded even when equal account is taken of the induction of genetic defects in all subsequent generations as well as that of fatal malignancies, and when the risks of accidental death in the occupation are also included. Thus in a typical occupation with an annual exposure of 4 mSv, the genetic and the fatal cancer risks together would, on the basis of the values given in Table 11.1, be responsible for about 7 such effects per year in every 100 000 exposed. Accidental mortality is found to be low in the staff of nuclear plants, at between 1 and 2 deaths per year per 100 000, and is probably as low or lower in medical radiological work.

In some forms of work with radiation, however, not only are the annual exposures higher, but the associated accidental risks are high also and in themselves alone prevent the work from being regarded as having a high degree of safety. In uranium mining, for example, an average exposure of 30 mSv per year would correspond to an annual causation of about 50 deaths or genetic effects per 100 000 miners, but accidental mortalities of between 2 and 3 times this rate have been recorded in some conditions of underground mining. It is likely also that, in some types of industrial radiography, and especially perhaps in the field use of mobile sources, the somewhat raised radiation exposures will be associated with a substantially greater accidental mortality than in most forms of radiation work.

Just as different occupations vary widely in their risks of accidental death, therefore, so they also vary in their risks from radiation, although certainly not so greatly. The limit on the exposure of any worker in any year is effective in preventing the rates of severe radiation effects from approaching the high rates of accidental death that are seen in some conventional occupations. The need for all practicable reductions in radiation exposure at work, however, remains as important as the need to reduce every other hazard in the working environment.

Index of total occupational risk

It is obvious that the risks of different occupations can only be compared in an approximate and incomplete way by assessing the frequency of deaths, or of deaths and genetic effects, in each. Radiation causes cancers which can be cured as well as those which cannot, and genetic defects which cause minor, rather than major, hardship and impairment of health. Similarly most occupational accidents are not fatal but result in longer or shorter periods of disability and loss of working capacity, and the same is true of diseases of occupational origin. Must we not do better than simply to compare mortality rates?

The attempt to do so, and to develop some general index of the total occupational harm, presents a number of difficulties. It is bad enough to try to assess a relative importance of one accidental death tomorrow and one death from cancer in twenty years time; or one fatal cancer in the person exposed and one genetic defect in his great-grandson. It is harder still to compare the hurt and hardship from one cancer which proves fatal and one which is cured by prolonged treatment, or by a quick and effective operation; or to make any numerical comparison between a fatal accident and a number of non-fatal ones. And yet any serious attempt to compare the safety of different working conditions must work towards a way of assessing different aspects of harm, and looking at their relative degrees of hardship. Indeed, the very act of proposing 'weighting' factors for different types of disability or effect is likely to evoke a clearer opinion as to what weight people do in fact attach to these disabilities.

One useful first approach to this problem has been to assess the length of life, or of full health and activity, that is lost as a result of the various types of occupational injury or disease and the frequency with which they occur.[91] The problem of weighting still remains: what weight would you give to the loss of 30 years of life as a result of one premature accidental death, as compared with an equal total period of lost health or activity from a series of non-fatal accidents? In many industries the annual loss of working time from non-fatal accidents is about equal to the loss of years of expectation of life from such accidental deaths as occur in the same working population in the same period of time. In so far as I would rather be off work at home or in hospital for a month, than dead for a month, I would, at least in theory, attach more weight to the fatal than to the non-fatal accidents in such industries. At least the loss of life expectancy from fatal cancers, and the periods of illness from fatal and non-fatal cancers, can be reviewed in relation to the loss of life and the periods of illness and disability from accidents or disease in other occupations; and an assessment has been made for the genetic defects of the types induced by radiation, of the length of impaired life after the defect has developed, and the loss of life due to early death in some of these conditions.[23]

Estimates have been made in this way of the lengths of life lost or impaired annually per thousand workers in different industries from all forms of occupational injury and recognized disease. On the basis of one such published assessment,[91] which took account of all medical effects, fatal or non-fatal, an exposure of 4 mSv per year would involve a total detriment equivalent to that in an industry with an annual accidental death rate of 2 per 100 000. An exposure of 30 mSv per year would be equivalent to a rate of 18 per 100 000. Such assessments, however, are likely to be of less value for any simple, unequivocal index that they could produce of the relative safety or risk of different industries, than in eliciting or developing opinions on the relative weights that it would be appropriate to attach to different harmful effects, and in indicating what are felt to be the major contributors to the total harm.

CONSEQUENCES OF PUBLIC EXPOSURES TO RADIATION

In considering the importance of occupational exposures, our primary concern needed to be with the safety of the individual, although the genetic effects of these exposures are significant as contributing to the total genetic irradiation of the population as a whole.

When we review the effects of the radiation exposure of members of the public, the doses to individuals are ordinarily so much smaller, and the numbers of people exposed are so much larger, that the impact on the community of any particular source will usually merit as much attention as the effect on the individual of the small doses received from it.

Harm to the community as a whole

The amount of harm that may be being caused in the community as a whole by radiation from different sources can be estimated from the average doses received annually from these sources, and the probability of harm per unit dose. In doing so, however, it must be remembered that most public exposure is to radiation of low LET, and that for such radiation the probability that harm will result from low doses may be substantially overestimated by using the assessments of harm observed in epidemiological surveys at higher dose.

Table 11.5

Annual numbers of deaths from cancer and of inherited defects in UK (population 55 million), and estimated numbers attributable to different radiation sources (assuming induction rates as in Table 11.1)

	Fatal cancers		Inherited defects	
	Average dose (effective)	Number	Average dose (genetic)	Number
Total, all causes		130 000		10 000*
Radiation from	mSv/yr		mSv/yr	
Natural sources	1.86	1250	1.00	270
Medical exposures	0.50	330	0.12	32
Fallout	0.010	7	0.0065	1.8
Occupational exposure	0.009	6	0.0045	1.2
Nuclear wastes	0.003	2	0.0022	0.6
Miscellaneous sources	0.008	5	0.0040	1.1

*Autosomal, x-linked, and chromosomal diseases, of frequencies maintained by mutation; also about 25 000 congenital malformations and 28 000 other abnormalities of irregular or multifactorial inheritance.[23]

For the UK population of about 55 million, however, Table 11.5 indicates that, on the basis of these risk estimates (as given in Table 11.1, p. 166), radiation from natural sources may be causing about 1 per cent of all fatal cancers. All other forms of radiation exposure would cause rather less than 0.3 per cent of cancer deaths, with medical exposures causing 0.25 per cent and the four remaining sources together being responsible for 0.015 per cent, or 20 of the 130 000 such deaths occurring annually.

The proportion of all inherited defects that may be attributable to radiation can be estimated similarly, except that now the relevant doses are those delivered to the germinal tissues in people at ages prior to the conception of children. On this basis, natural sources of radiation may be causing some 270 inherited abnormalities each year, since, with a dose rate of 1 mSv per year to the germinal tissues, about 30 mSv will have been delivered to these tissues by the average age at conception of children. With the risk of 15 induced defects per million children per millisievert received prior to conception, 270 such defects would result among the 600 000 children born annually. This number of defects may be compared with the 10 000 children born each year with an inherited disease or defect attributed to mutations from all causes, or with the substantially larger numbers with inherited diseases or anomalies of which the frequency is not regarded as being maintained by mutation.[23]

Other sources of radiation exposure would, if they, like the natural sources, had continued for many generations, add a further 35 to 40 defects annually, with rather over 30 from medical exposures and an additional 5 from fallout, occupational exposures, nuclear wastes, and miscellaneous sources together. (The contribution from nuclear wastes would be increased from 0.6, as in Table 11.5, to about 2 by inclusion of exposures due to possible releases in major accidents or from high activity wastes, on the basis indicated in Table 6.2 and in the notes to Table 6.4.)

The estimation of the occupational risks of radiation exposure made it possible to compare these risks with those occurring in other occupations, and so to form some opinion of the relative safety of different types of work. In much the same way, estimates have been made of the total amount of harm caused within the community by generating equal amounts of electricity from different primary fuel sources.[92–94] For example, if the output of one gigawatt of electricity for one year from nuclear sources results in a total collective dose of 100 man-sieverts from all phases of the nuclear fuel cycle, the supply of the present electricity requirements of one or two million people for a year from this source would cause about one fatal cancer and one induced genetic defect. Accidental fatalities, largely in mining the uranium, would add rather less than one death to this total. This estimate, of between two and three deaths or substantial genetic effects, has been compared with the number of deaths that are likely to be caused in the supply of an equal amount of electricity from other sources: with estimates of less than one death per gigawatt year derived from natural gas, and of between three and six for hydroelectric supplies. With coal as the primary fuel, the total harm is probably rather greater, since fatal accidents in mining and in transport alone account for between two and three deaths per GW year output, while lung disease in miners may be responsible for twice as many, and the environmental effects of discharges from coal fired power plants probably cause a greater number still, although the harmful effect of these discharges is much less well quantified than that from radioactive effluents.

Although these estimates of the harmfulness of different alternative sources vary widely, from less than one to over ten deaths for equal output of electricity, even the highest estimates of these annual risks of power production are small compared to the number of deaths occurring naturally each year, of from 12 000 to 24 000, in the populations of one or two million supplied with their electricity at the expense of these risks; and, of course, many other factors necessarily affect the choice of fuels used. The amount of harm caused, however, although small, ought surely to have some influence on decisions, and its estimation must therefore be of importance; and the disturbingly large differences in electrical power available in different parts of the world (Table 6.3, p. 86) emphasize the need to consider the safety of all available sources.

Harm caused to individuals

It is easy to assess the average risk to an individual from present rates of radiation exposure, since this of course follows from the estimates for the population as a whole: that all sources of exposure may be responsible for in the region of 1 per cent of his or her risk of fatal cancer or of transmitting an inherited abnormality. It is also easy to see that the risks will vary according to the different amounts of radiation that different people may receive. But is the risk the same for different individuals who have received equal doses? Certainly not when the doses are received at different ages, or in people of different sexes. But apart from these differences, do individuals vary fundamentally in their susceptibility to harm by radiation?

Variation in individual susceptibility

We know that, if a thousand people are each exposed to a high dose of one sievert, a fatal cancer is likely to be induced in about twelve of them; but we do not know which twelve, or whether there would be any way of detecting who the unfortunate twelve would be. We know that, in the development of a cancer, many of the cells formed by cell division in the cancer clone do not survive, and the development of a cancer depends upon the few which do survive. There could therefore, for example, be individual differences in this cancer cell killing process which were independent of radiation, but which influenced the frequency of its harmful effects.

It is also known that, in a certain rare hereditary disease, the cells are unusually sensitive to the cell killing effects of radiation, although not correspondingly sensitive to its effects in causing mutations.[23] People suffering from this disease have a high risk of certain forms of lymphatic cancer. Even if they were sensitive to induction by radiation of cancer of all types, however, there is no evidence to suggest that they, or people who carried their (recessive) type of gene heterozygously, could constitute the minority of individuals who develop cancer following radiation exposure, a minority which in any case varies with the size of the dose rather than representing a fixed sensitive proportion of the population.

While, therefore, there continues to be no good evidence of any subgroups of the population who are specially sensitive to the carcinogenic or mutagenic effects of radiation, or the possible size of any such subgroups or how they might be detected, the risk to individuals can best be expressed on a basis of the doses that they receive, and the average risks as determined in the population as a whole.

The members of the community who provide an exception to this statement and who do require special care and attention are the unborn children developing *in utero*. Not only does the risk of cancer induction appear to be somewhat greater than in adults (p. 140), perhaps by a factor of 2 or 3, following radiation received during the latter part of pregnancy. In addition, developmental abnormalities may be caused if relatively low threshold doses are exceeded at earlier stages in pregnancy when organ development is taking place.[95]

This risk is likely to be very small indeed before the existence of a pregnancy can be recognized, by a normal menstrual period having been missed. During the following month, no such effects are to be expected unless substantial doses, probably of several tens of millisieverts, are received by the embryo during any of the short time prериods, of a few days only, during which different organs or tissues are developing. Even large annual exposures, therefore, if received at a roughly constant dose rate, would not be likely to cause any threshold levels to be exceeded. During the time while the brain is developing, however, the evidence from Hiroshima and Nagasaki has indicated a longer period of sensitivity, of about seven weeks in mid pregnancy, during which exposures of some tens of millisieverts or more were followed by impaired mental development.

Risk from natural sources

The natural sources of radiation deliver similar annual doses to most members of the population, and few people are likely to receive more than 0.2 Sv from them during their lifetime in most countries. In the UK it would be rare for these sources to add more than about 0.5 per cent to the 22 per cent risk of dying from cancer due to other causes, even under conditions of increased regional radioactivity.

Risk from medical procedures

The risks to individuals from the radiological examinations that they incur are ordinarily very small. It has been mentioned above (p. 138) that no evidence of harm has been detectable, except following the use of Thorotrast or irradiation *in utero*, unless some hundred or more chest x rays had been made at rather high dose by fluoroscopy on each of a large number of patients. The typical (median) effective dose from a diagnostic x ray is of rather less than one millisievert, and that from a radionuclide investigation has about half this value.[96] It would therefore require over a thousand such examinations to cause a 1 per cent risk of a fatal cancer induction.

In radiotherapy, where the local tissue doses are considerably higher, the risks are less trivial, and a fatal cancer risk of 1 per cent has been noted as following radiation treatments of ankylosing spondylitis. Since radiotherapy is commonly used in the control of malignant disease for which other forms of treatment are inappropriate or inadequate, even risks of this size may commonly appear to be warranted. In less life-threatening conditions, however, radiotherapy at high dose is used decreasingly in many countries. One of the essential values of obtaining even an approximate estimate of the overall risks of radiation, or of any other agent used in treatment, however, must be to help in ensuring that the risk involved in the treatment proposed is likely to be less than that for any equally available alternative treatment, and certainly less than that of omitting any active treatment at all.

Risk from fallout

The fallout from past atmospheric weapon tests will irradiate the world's population for very many generations at a very low rate, mainly from the long-lived carbon-14 remaining in the environment. The size of the total collective dose from these tests depends therefore on the future size of the world population that will be so exposed. If the population remained at its present four billion, the collective dose would be about 13 million man-sieverts. If it increased to ten billion, the collective dose could have about twice this value. It is impossible to estimate realistically the harm that might result over thousands of years from a single decade of heavy atmospheric testing in view of the uncertainties, not only in the growth of the world population, but also in the extent to which the risk estimates, of 12 fatal cancers per thousand man-sieverts, derived from high doses may overestimate such risks as might apply to exceedingly small increases above the radiation received from natural sources. Nor can we reasonably predict the reductions of risk that could result within a hundred or a thousand years from the prevention or cure of cancers and perhaps of genetic defects, or even from techniques for reducing the environmental carbon-14.

With no individual likely to receive more than about 2 mSv in his lifetime from global fallout, however, except in the arctic regions referred to above (p. 76), the greatest increase in his risk of death from cancer would be of about one in 40 000.

Risk from nuclear power production

To most individuals, the risks from nuclear power production will correspond to those as averaged over the community as a whole, with total effective doses received during any person's lifetime depending on the extent of power production from nuclear sources, but being unlikely to exceed about 0.5 millisieverts. (This perhaps rather extreme estimate is based on a world-wide electricity consumption of 1 kilowatt per person, 50 per cent of it being derived from nuclear sources, and delivering 11 man-Sv per GW(e)y to current generations of members of the public, as in Table 6.2). This could impose a six in a million risk of fatal cancer development.

We need, however, to look at two particular situations: firstly, that of any sections of the public who are more highly exposed than the average, because of local or other factors, and secondly, individuals exposed as a result of accidental releases from nuclear plants.

'Critical groups' of the public The exposure of members of the public to radiation clearly requires to be limited in the same way as should apply to occupational exposures, but presumably with a lower limit, since occupational exposure does not involve children, is associated with individual monitoring of doses and with medical surveillance, and, when at an appropriately low level, could be regarded as a component of the element of risk which is common to all occupations.

In these circumstances, the International Commission on Radiological Protection has recommended[30] that no members of the public should receive more than one tenth (5 mSv) of the annual occupational dose limit in any year, as the total exposure from all sources other than natural ones and those from necessary medical procedures, which clearly should not be restricted in this way. It also suggested that in conditions in which such a 'critical' group of the population might receive prolonged or lifelong exposures in excess of natural and medical radiation, the average annual rate should not be expected to exceed one fiftieth of the occupational limit, or 1 mSv per year.

Few instances have been identified in which these limits might be being approached, and none in which they have been exceeded. The maximum likely dose rates to local communities due to aqueous discharges from reactors and reprocessing plants in the UK are estimated and reported annually by the Fisheries Radiobiology Laboratory at Lowestoft.[78] These rates have in most cases been low, at a few per cent of the 5 mSv annual limit. The doses resulting from discharge of radioactive caesium to the sea from reprocessing plant at Windscale, however, were estimated to have risen from below 1 mSv per year prior to 1975 to a value of 2.2 mSv in 1976 in the most exposed groups eating locally caught fish at Whitehaven, falling again to an average of 1.2 mSv per year during the period from 1978 to 1980.[78]

These assessments were based on the measured radioactivity of locally caught fish and molluscs and estimates of their consumption. They were however rather accurately confirmed when a group of 17 people living at Whitehaven volunteered, during the Windscale Inquiry in 1977, to have measurements made by whole body counters after weighing and eating the amount of fish they normally consumed.[97]

The estimated risk of fatal cancer in any individual who actually received a lifelong dose rate of 1 mSv per year would be rather less than 1 in 1000, increasing the normal risk of death from this disease by about ½ per cent of its value. Such an increase would not be detectable statistically in any small group of people who were locally exposed, and is within regional variations in cancer incidence anyhow. It is clearly important therefore to estimate the likelihood and amount of any such local exposures, and to limit their frequency as much as is practicable, even though the total annual exposure is within the range of variation of that from natural sources.

Major accidents Several detailed estimates have been made of the number of deaths, and of other harmful effects on the individual and on the community, that might result from the most severe types of possible reactor accidents,[98] and of the probability that any such accident would occur, for example during every million reactor-years of operation. The estimated numbers of deaths ordinarily range up to figures similar to those recorded for dam failures,[92] and are higher by a factor of about ten than those from individual marine accidents.[99] The likelihood of such severe reactor accidents however is estimated to be very small.

The effects that severe types of reactor accident would have, if they occurred, can be estimated more reliably than the probability that they might occur. A recent study has examined in considerable detail the possible consequences of 12 types of accident, all being in the severe categories resulting from overheating and melting in the reactor core with release of radionuclides to the environment. In the five which involved the most severe consequences, the most probable numbers of early deaths ranged from 10 to 130, and the numbers of fatal cancers developing later from 1400 to 3600.[98] These figures were based on earlier (1975) estimates of the size of releases from pressurized water reactors following melt-down in the core. More recent (1982) evidence on retention of activity in the primary cooling circuit, its removal from the atmosphere in containment structures, and during leakage from the containment, gave rather lower figures, with few early deaths and from 400 to 1300 late fatal cancers.

The number of such effects, and of the other harmful biological and environmental effects analysed, were shown to vary widely with such factors as wind direction and weather conditions at the time of the supposed accident, the current patterns of land use in food production, and the counter measures adopted following the release; and the severity of the effects will vary with reactor site and the population distribution round it. These differing probabilities combine with the estimated probability of different types of reactor accident to give some estimate of the number of fatalities that might occur during long periods of operation of large numbers of reactors. The study made in the Federal German Republic[50] was based on the population and meteorological conditions around 25 reactor sites, and suggested the occurrence of 200 early fatalities in any 25 million reactor-years of operation, and 2700 late deaths from cancer in 25 thousand reactor-years. Despite the severe effects that some forms of reactor accident would have if they occurred, therefore, the probability of their occurrence, and so the lifetime risk to any member of a neighbouring community, would appear likely to be very small. For example, if one reactor of the output considered in the German studies was supplying the electricity needs of a million people, an average rate of 27 deaths in this community in every 250 years would correspond to an annual risk to an individual in that community of one in nine million.

RISKS AND RISK PERCEPTION

The impact of occasional major disasters, and the anxiety caused by the possibility of their occurrence, is of course much greater than that from the same number of deaths occurring singly at a regular rate during an equal period of time. In the USA, major disasters, such as from fires, floods, or storms which caused 30 or more deaths at a time, involved an annual death rate of 385 in the period from 1900 to 1965. Over a similar period (1903-65) the annual death rate[88, 90] from all accidents averaged 89 500. Yet the disasters will certainly have received much greater notice, and will probably have caused greater anxiety and distress, than the 200 times greater toll of deaths occurring singly.

It does seem important, however, to obtain firm numerical evidence of the size of the risks to individuals or to the community from different causes: not as any sole or overriding criterion for decisions, because the nature of the risk and the way it is imposed will always affect our attitude towards it; but as giving an added perspective to its importance or its acceptability. It must be an oversimplification to regard some actions or situations as safe, and others as unsafe, when a greater or less degree of safety or risk attaches to all human activities (and it has been said that the word 'safe' should never be allowed out unless accompanied by a responsible adverb). Certainly in some situations the risk may be so trivial as to be properly ignored, or so large that it is overriding. Too often, however, the risks of harm resulting from a proposed action are simply not recognized or not known. It is valuable that recent attention to the assessment of risk[89, 96, 100-103] has improved our knowledge not only on the actual size of a variety of risks, but also increasingly on the factors which influence our assessment of the importance to us of various kinds of risk – whether they are self chosen or sought, as in many forms of sport, or generally accepted, as in risks from most medical procedures or modes of travel, or are imposed on us by the hand of nature or of man.[89]

It is a useful aspect of the studies in radiation dosimetry, biology, and epidemiology that we can now go some way towards assessing objectively the risk from different sources of radiation exposure, both in an absolute sense, and relative to each other and to other risks of life. It would be valuable if the safety or risk of many other environmental agents were assessed in the same way.

REFERENCES

Note. Certain reports or periodical publications quoted below are published or issued from the following addresses:

Annals of the ICRP: Pergamon Press, Headington Hill Hall, Oxford.
Commission of the European Communities: through HMSO, 49 High Holborn, London WC1.
Fisheries Radiobiology Laboratory: Directorate of Fisheries Research, Fisheries Laboratory, Lowestoft, Suffolk.
International Atomic Energy Agency, Division of Publications, Wagramerstrasse, PO Box 100, A-1400 Vienna; or through HMSO.
National Radiological Protection Board, Chilton, Didcot, Oxfordshire.
US Government reports: through NTIS Microinformation, PO Box 3, Newman Lane, Alton, Hampshire.
UNSCEAR: through HMSO.
WHO: through HMSO.

1. Seelmann-Eggebert, W., Pfennig, G., and Münzel, H. *Nuclid Karte.* Gersbach u. Sohn Verlag, Munich (1974).
2. Silk, J. *The big bang: the creation and evolution of the universe.* Freeman, San Francisco (1980).
3. Beck, C. *Roentgen ray diagnosis and therapy.* Appleton, New York (1904).
4. Taylor, L.S. *Organisation for radiation protection.* Office of Health and Environmental Research and US Department of Energy. Washington DC (1979).
5. Hevesy, G. *Radioactive indicators.* Interscience, New York (1948).
6. Parker, R.P., Smith, P.H.S., and Taylor, D.M. *Basic science of nuclear medicine.* Churchill Livingstone, Edinburgh (1978).
7. Maisey, M. *Nuclear medicine, a clinical introduction.* Update, London (1980).
8. Gottschalk, A. and Potchen, E.J. *Diagnostic nuclear medicine.* Williams & Wilkins, Baltimore (1976).
9. Rothfeld, B. (ed.) *Nuclear Medicine in vitro.* Lippincote, Philadelphia (1974).
10. *Medical radionuclide imaging 1980* (2 vols.). International Atomic Energy Agency, Vienna (1981).
11. *Diagnostic imaging. Br. Med. Bull.* **36**, 205–41 (1980).
12. *Isotopes in day to day life.* International Atomic Energy Agency, Vienna (1977).
13. International Atomic Energy Agency Bulletin, Vol. 20 Part 5 (1978); Vol. 23 Part 3 (1982).
14. *Report of the United Nations Scientific Committee on the effects of atomic radiation.* General Assembly official records: 17th session, Supplement No. 16 (A/5216). United Nations, New York (1962).
15. Fleming, S.J. *Authenticity in art: the scientific detection of forgery.* The Institute of Physics, London (1975).

16. Wiltshire, W.J. *A further handbook of industrial radiography*. Edward Arnold, London (1957).
17. Offori, E.D. The scourge of the tsetse. *Int. At. En. Agency Bull.* **23**, 43–6 (1982).
18. Smyth, H.D. *Atomic Energy*. US Government Printing Office (1945), reprinted HMSO London (1945).
19. Takeshita, K. Dose estimation from residual and fallout radioactivity. Areal surveys. *J. Radiat. Res.* **16** (suppl.) 24-31 (1975).
20. Okajima, S. *Dose estimation from residual and fallout radioactivity. Fallout in the Nagasaki – Nishiyama district. J. Radiat. Res.* **16** (suppl.) 35-41 (1975).
21. Okada, S., Hamilton, H.B., Egami, N., Okajima, S., Russell, W.J., and Takeshita, K. (eds.) Preface to *J. Radiat. Res.* **16** (Suppl.) i–iii (1975).
22. Beebe, G.W., Kato, H., and Land, C.E. *Mortality experience of atomic bomb survivors, 1950-1974.* Life span study report 8, RERF Technical Report TR 1-77. Radiation Effects Research Foundation (1977).
23. *Ionizing radiation: sources and biological effects.* United Nations scientific committee on the effects of atomic radiation. 1982 report to the General Assembly (Sales No. E82.IX.8). United Nations, New York (1982).
24. *Nuclear power, the environment and man.* Information booklet prepared by the International Atomic Energy Agency and the World Health Organization. IAEA, Vienna (1982).
25. *Health implications of nuclear power production.* World Health Organization regional office for Europe (WHO regional publications, European series No. 3), Copenhagen (1978).
26. Agricola, G. (Georg Bauer). *De re metallica* (1556).
27. Cowan, G.A. A natural fission reactor. *Scient. Am.* **235**, 36-47 (1976).
28. Naunet, R. The Oklo phenomenon. *International Atomic Energy Agency Bull.* **17**, 22-4 (1975); and see *idem* **17**, 2-4 and 44-5 (1975).
29. *Sources and effects of ionizing radiation.* United Nations scientific committee on the effects of atomic radiation. 1977 report to the General Assembly (Sales No. E77. IX. 1). United Nations, New York (1977).
30. *Recommendations of the International Commission on Radiological Protection.* ICRP Publication 26. *Annals of the ICRP* **1** (3), 1-53 (1977).
31. McKinlay, A.F. *Thermoluminescence dosimetry*. Adam Hilger, Bristol (1981).
32. *Limits of intakes of radionuclides for workers.* ICRP Publication 30, Parts 1, 2, and 3. *Annals of the ICRP* **2** (3/4), 1-116 (1979), **4** (3/4), 1-71 (1980), and 6 (2/3), 1-124 (1981).
33. Kelly, G.N., Jones, J.A., Bryant, P.M., and Morley, F. The predicted exposure of the population of the European Community resulting from discharges of krypton-85, tritium, carbon-14 and iodine-129 from the nuclear power industry to the year 2000. Commission of the European Communities (Doc. V/2676/75). Luxembourg (1975).
34. *Methodology for evaluating the radiological consequences of radioactive effluents released in normal operations.* Commission of the European Communities (Doc. V/3865/79), Luxembourg (1979).
35. Cullen, T.L. and Penna Franca, E. (eds.) *International symposium on areas on high natural radioactivity.* Poços de Caldas, Brazil 1975. Academia Brasileira de Ciências, Rio de Janeiro (1977).
36. Darby, S.C., Kendall, G.M., Rae, S., and Wall, B.F. *The genetically significant*

dose from diagnostic radiology in Great Britain in 1977. National Radiological Protection Board report NRPB-R106, Chilton (1980).

37. *Frequency of various medical procedures involving the use of ionizing radiation, 1978-1979.* World Health Organization, Geneva (1980).
38. Hashizume, T., Maruyama, T., Yamaguchi, H., Nishikawa, K., and Tateno, Y. Estimation of population doses from medical uses of radiopharmaceuticals in Japan, 1977. *Nippon Acta Radiol.* **39**, 747-60 (1979).
39. Peirson, D.H. and Cambray, R.S. Fission product fall-out from the nuclear explosions of 1961 and 1962. *Nature, Lond.* **205**, 433-40 (1965).
40. Hanson, W.C. Caesium-137 concentrations in North Alaskan Eskimos 1962-79: effects of ecological, cultural and political factors. *Hlth. Phys.* **42**, 433-47 (1982).
41. Conard, R.A. *et al. A twenty year review of medical findings in a Marshallese population accidentally exposed to radioactive fallout.* Brookhaven National Laboratory report (BNL 50424), Upton, NY (1975).
42. Nishiwaki, Y. Bikini ash. *Atomic Scientists Journal* **4**, 1-13 (1954).
43. Frank, A.L. and Benton, E.V. Measurements of gamma exposures in uranium mines. *Hlth. Phys.* **40**, 240-3 (1981).
44. Uzunov, I., Steinhäusler, F., and Pohl, E. Carcinogenic risks of exposure to radon daughters associated with radon spas. *Hlth. Phys.* **41**, 807-13 (1981).
45. Taylor, F.E. and Webb, G.A.M. *Radiation exposure of the UK population* National Radiological Protection Board report NRPB-R77, Chilton (1978).
46. Kelly, G.N., Jones, J.A., and Broomfield, M. *The radiation exposure of the UK population from airborne effluents discharged from civil nuclear installations in the UK in 1978.* National Radiological Protection Board report NRPB-R118, Chilton (1981).
47. Camplin, W.C., Clark, M.J., and Delow, C.E. *The radiation exposure of the UK population from liquid effluents discharged from civil nuclear installations in the UK in 1978.* National Radiological Protection Board report NRPB-R119, Chilton (1982).
48. Crick, M.J. and Linsley, G.S. *An assessment of the radiological impact of the Windscale reactor fire, October 1957.* National Radiological Protection Board report NRPB-R135, Chilton (1982).
49. US Regulatory Commission. *An assessment of the accident risks in US commercial nuclear power plants* (report WASH-1400). Nuclear Regulatory Commission, Washington, DC (1975).
50. Bundesministeriums für Forschung und Technologie, Deutsche Risikostudie Kernkraftwerke. *Eine Untersuchung zu dem durch Störfälle in Kernkraft werken verursachten Risiko.* Verlag TUV Reinland (1980).
51. *Effektivare Bredskap*, Vol. 1. Statens strålskyddsinstitut, Stockholm (1979).
52. International Atomic Energy Agency, reference data series No. 1 (September 1982). IAEA, Vienna (1982).
53. *Safeguards for geologic repositories.* International Fuel Cycle Evaluation working group report (INFCE/DEP/WG7/18). IAEA, Vienna (1979).
54. Rutherford, E., Chadwick, J., and Ellis, C.D. *Radiations from radioactive substances.* Cambridge University Press, (1930, reissued 1951).
55. R. Martinez, G., G. Cassab H, G. Ganem G., E. Guttman K., M. Lieberman L., H. Rodríguez M., and L. Vater B. Observaciones sobre la exposicion accidental de una familia a una fuente de cobalto-60. *Revta Méd. Inst. Mex. Seguro Soc.* **3** (suppl. 1) 13-68 (1964).
56. Ye Gen-yao, Liu Yong, Tien Nue, Chiang Ben-yun, Chien Feng-wei, and Xiae Chien-ling. The People's Republic of China accident in 1963. In *The*

medical basis for radiation accident preparedness (eds. K.F. Hübner, and S.A. Fry) pp. 81–9. Elvesier North-Holland (1980).

57. Balner, H. *Bone marrow transplantation and other treatment after radiation injury*. Commission of the European Communities report (EUR 5884 e). Martinus Nijhoff, The Hague (1977).
58. Rubin, P. and Casarett, G.W. *Clinical radiation pathology*. Saunders, Philadelphia (1968).
59. Larsen, P.R., Conard, R.A., Knudsen, K.D., Robbins, J., Wolff, J., Rall, J.E., Nicoloff, J.T., and Dobyns, B.M. Thyroid hypofunction after exposure to fallout from a hydrogen bomb explosion. *J. Am. Med. Ass.* **247**, 1571–5 (1982).
60. Evans, R.D., Keane, A.T., Kolenkow, R.J., Neal, W.R., and Shanahan, M.M. Radiogenic tumors in the radium and mesothorium cases studied at MIT. In *Delayed effects of bone-seeking radionuclides* (ed. C.W. Mays, W.S.S. Jee, R.D. Lloyd, B.J. Stover, J.H. Dougherty, and G.N. Taylor) pp. 157–94 University of Utah Press, Salt Lake City (1969).
61. Wei Luxin *et al.* Health survey in high background areas in China. *Science N.Y.* **209**, 877–80 (1980).
62. Sundaram, K.; Edwards, J.H. and Harnden, D.G.; Verma, I.C., Kochupillai, N., Grewal, M.S., Ramachandran, K., and Ramalingaswami, K. Down's syndrome in Kerala. *Nature, Lond.* **267**, 728–9 (1977).
63. Evans, H.J., Buckton, K.E., Hamilton, G.E., and Carothers, A. Radiation-induced chromosome aberrations in nuclear-dockyard workers. *Nature, Lond.* **277**, 531–4 (1979).
64. Lloyd, D.C., Purrott, R.J., and Reeder, E.J. The incidence of unstable chromosome aberrations in peripheral blood lymphocytes from unirradiated and occupationally exposed people. *Mutat. Res.* **72**, 523–32 (1980).
65. Searle, A.G. Use of doubling doses for the estimation of genetic risks In *Proc. European symposium on rad equivalence, Orsay 1976*, pp. 133–45. Commission of the European Communities report (EUR 5725 e), Luxembourg (1977).
66. Schull, W.J., Otake, M., and Neal, J.V. Genetic effects of the atomic bombs: a reappraisal. *Science, N.Y.* **213**, 1220–7 (1981).
67. Symington, T. and Carter, R.L. (eds.) *Scientific foundations of oncology*. Heinemann, London (1976).
68. Swent, L.W. Statement of principles. In *Hazards in mining*, International conference; Golden, Colorado, October 1981. M. Gomez (ed.), pp. 4–7. American Institute of Mining, Metallurgical and Petroleum Engineers, New York (1981).
69. Protection against ionizing radiation from external sources used in medicine. ICRP Publication 33. *Annals of the ICRP* **9** (1), 1–69 (1982).
70. Protection of the patient in diagnostic radiology. ICRP Publication 34. *Annals of the ICRP* **9** (3/4), 1–82 (1982).
71. *Protection of the patient in radionuclide investigations*. ICRP Publication 17. Pergamon, Oxford (1971).
72. *Factors influencing the choice and use of radionuclides in nuclear medicine*. (US) National Council on Radiation Protection and Measurements, NCRP Report No. 70. Bethesda, (1982).
73. The handling, storage, use and disposal of unsealed radionuclides in hospitals and medical research establishments. ICRP Publication 25. *Annals of the ICRP* **1** (2), 1–46 (1977).

74. General principles of monitoring for radiation protection of workers. ICRP Publication 35. *Annals of the ICRP* **9** (4), 1-36 (1982).
75. Radiation protection in uranium and other mines. ICRP Publication 24. *Annals of the ICRP* **1** (1), 1-28 (1977).
76. Limits for inhalation of radon daughters by workers. ICRP Publication 32. *Annals of the ICRP* **6** (1), 1-24 (1981).
77. Cronquist, A., Pochin, E.E., and Thompson, B.D. The speed of suppression by iodate of thyroid iodine uptake. *Hlth. Phys.* **21**, 393-4 (1971).
78. *Radioactivity in surface and coastal waters of the British Isles.* (1963 to 1980). Ministry of Agriculture, Fisheries and Food: Fisheries Radiobiological Laboratory reports (Technical reports FRLI of 1967, to Aquatic Environmental Monitoring report No. 8 of 1982). Lowestoft (1967-82).
79. Morris, B. and Marples, A. See-through solution to nuclear waste. *New Scient.* **83**, 276-8 (1979).
80. Ringwood, T. Safety in depth for nuclear waste disposal. *New Scient.* **88**, 574-5 (1980).
81. *The management of radioactive wastes* (81-3652). International Atomic Energy Agency, Vienna (1981).
82. Cohen, B.L. Health effects of radon emissions from uranium mill tailings. *Hlth. Phys.* **42**, 695-702 (1982).
83. Zettwoog, P., Fourcade, N., Campbell, F.E., Caplan, H., and Haile, J. The radon concentration profile and the flux from a pilot-scale layer tailings pile. *Hlth. Phys.* **43**, 428-33 (1982).
84. *The hazards to man of nuclear and allied radiations.* Medical Research Council (Cmd 9780). HMSO, London (1956).
85. *The hazards to man of nuclear and allied radiations. A second report.* Medical Research Council (Cmd 1225). HMSO, London (1960).
86. *The effects on populations of exposure to low levels of ionizing radiation: 1980* (the 'BEIR III report'). National Academy Press, Washington DC (1980).
87. Bond, V.P. Radiobiological input to radiation protection standards. *Hlth. Phys.* **41**, 799-806 (1981).
88. Pochin, E.E. Occupational and other fatality rates. *Commun. Health* **6**, 2-13 (1974).
89. *Risk assessment: a study group report.* Royal Society, London (1983).
90. *Accident facts.* US National Safety Council, Chicago (1956-81).
91. Problems involved in developing an index of harm. ICRP Publication 27. *Annals of the ICRP* **1** (4), 1-24 (1977).
92. Inhaber, H. *Risk of energy production.* Atomic Energy Control Board of Canada report AECB-1119 (Rev. 3), Ottawa (1980).
93. Pochin, E.E. Biological risk involved in power production. *Phys. Technol.* **11**, 93-8, 110 (1980).
94. The respective risks of different energy sources: report of symposium, Paris, January 1980. *Int. At. En. Ag. Bull.* **22**, 35-128 (1980).
95. Russell, L.B. Irradiation damage to the embryo, fetus and neonate. In *Biological risks of medical irradiations*, (ed. G.D. Fullerton, Kopp, D.T., Waggener, R.G., and Webster, E.W.) pp. 33-54. American Institute of Physics, New York (1980).
96. Warner, F. (ed.) *The assessment and perception of risk: a Royal Society discussion.* Royal Society, London (1981).

97. Mr Justice Parker. *Report of the Windscale inquiry.* HMSO, London (1978).
98. Kelly, G.N. and Clarke, R.H. *An assessment of the radiological consequences of releases from degraded core accidents for the Sizewell PWR.* National Radiological Protection Board report NRPB-R137, Chilton (1982).
99. Fryer, L.S. and Griffiths, R.F. *Worldwide data on the incidence of multiple-fatality accidents.* UK Atomic Energy Authority, Safety and Reliability Directorate report SRD R149, Warrington (1979).
100. Whyte, A.V. and Burton, I. (ed.) *Environmental risk assessment.* Scientific Committee on Problems of the Environment report SCOPE 15 Wiley, Chichester (1980).
101. Berg, G.G. and Maillie, H.D. (ed.) *Measurement of risks.* Plenum, New York (1981).
102. Pochin, E.E. The need to estimate risks. *Phys. Med. Biol.* **25**, 1–12, (1980).
103. Pochin, E.E. Risk assessment in relation to medical ethics. *J. Med. Ethics* **8**, 180–4 (1982).

GLOSSARY

See Index (at pages shown in bold type) for words of which the meaning is discussed in the text.

Aberrations, chromosomal. Abnormalities in number or structure of chromosomes.

Activation analysis. Determination of chemical composition by rendering certain types of atom radioactive, e.g. by neutron irradiation.

Activity, of radioactive substance. Measure of number of atoms disintegrating per second in the sample. Unit: one becquerel (Bq) corresponding to one disintegration per second.

Acute effects, e.g. of radiation. Effects occurring within days or weeks of radiation exposure.

Amino-acids. Simple chemical compounds of which proteins are composed.

Aneurism. Dilatation of an artery due to weakening of its wall.

Ankylosing spondylitis. Disease causing reduction or loss of movement of spinal and other joints.

Antibodies. Chemical compounds formed by the tissues maintaining body immunity, e.g. to combine with and neutralize harmful substances.

Banding of chromosomes. Successive transverse bands of material in chromosomes, differentiated by staining and corresponding to different aspects or phases of chromosome activity.

Bases in DNA structure. Chemical compounds of four kinds (adenine, cytosine, guanine, or thymine) attached in sequence along the DNA molecule and determining the coding information of genes.

Becquerel (Bq). Unit of radioactivity (p. 68).

Benign, as contrasted with malignant, tumour or neoplasm. *See* Cancer.

Biosphere. Region of earth occupied by animal or vegetable life.

Calcified tissues. Bone, or other tissues in healed or active disease, having a high calcium content.

Calorimetry. Measurement of quantity of heat.

Cancer and related terms. Cancer is commonly used in the text to refer to all forms of malignant growth or tumour, including those of bone marrow (leukaemia), bone (osteosarcoma), or carcinoma. Cancer mortality: frequency of deaths from cancer. Cancer incidence: frequency of occurrence of cases of cancer (whether fatal or not).

Carcinogenic. Causing cancer.

Chelates, chelating agents. Substances to which atoms, e.g. of plutonium may become attached, facilitating their excretion from the body.

Child expectancy. The average number of children likely to be conceived subsequently, by an individual of stated age and sex.

Clone. A population of cells all derived by successive cell divisions from one original cell.

Co-precipitation. The precipitation from solution of one chemical substance, achieved by causing the precipitation of another similar substance.

Critical group of individuals. The group identified, on grounds of diet, locality, age, etc., as likely to be the most exposed to radiation from a particular source.

Criticality, critical mass. Criticality occurs when a large enough mass (a 'critical mass') of a fissile material accumulates within a small enough volume to 'become critical' and begin to start a chain reaction of fissions.

Crookes tube. A sealed glass tube from which most of the air has been removed, and through which an electric current can be passed.

Crust of the earth. The surface layer to a depth of about 20 km below continents, or 8 km below the ocean floors.

Crystallography. Determination of the structure and dimensions of molecules by x-ray methods.

CT, or CAT, scanning. Scanning of body structures (by computerized axial tomography), with narrow beams of x rays passing through each section along the body axis from different directions.

Curie (Ci). The former unit of (radio-) activity. One curie of any sample of radioactive material underwent the same number of atomic disintegrations per unit time as occurred in one gram of radium-226. Replaced by the becquerel as SI unit (with 1 Ci corresponding to 3.7×10^{10} Bq).

Curve stripping. An electronic method of measuring the contributions to a curve that are due to minor contributors, by successively subtracting those due to the major contributors.

Cyclotron. An electrical device for accelerating charged particles in a beam directed at a target, to use or study the effect of the particles on the target material.

Cytoplasm. The part of a cell surrounding the nucleus, and bounded by the surface membrane of the cell.

Decay chain. The sequence of radioactive atoms produced by successive decay from an original or primordial radionuclide (p. 8), and terminating when a stable form of atom is reached.

Dicentric chromosomes. Abnormal chromosome pairs having two, instead of the normal one, 'centre' at which the two chromosomes of the pair become joined prior to cell division.

Disintegration of the nucleus of a radioactive atom. Occurs with emission of radiation, and decay to a new nuclear form.

DNA. Deoxyribonucleic acid (p. 112).

Dose, of radiation. When not otherwise qualified, refers to the effective dose equivalent (p. 48).

Dose, fractionated. A total dose delivered in successive fractions, with periods of days or weeks between each fraction. Contrast with protracted doses given by irradiation continuously over a long period.

Electroscope. Device for measuring electrical charge.

Endosteal cells. Bone cells located on the inner surfaces of bone, from which most bone cancers develop.

Enrichment of uranium. Increase in the concentration in uranium of its fissile isotope uranium-235.

Enzymes. Types of protein which accelerate or make possible certain biochemical processes.

Epidemiology. The study of the frequency of disease or abnormality, usually in human populations.

Exposure to radiation. Used in the sense of receiving a dose of radiation, or of the dose received.

Field, x-ray. The volume exposed to the x-ray beam, in diagnostic or therapeutic radiology.

Fibrosis. The development of fibrous tissue following damage to body tissues, which become 'fibrotic'.

Fissile or fissionable atomic nuclei. Ones liable to become split (or fissioned), e.g. when irradiated with neutrons, resulting in fission products, or fission fragments, of smaller nuclear mass.

Flux, e.g. of neutrons. Number of neutrons crossing unit area in unit time.

Fly ash. Component of ash from combustion of fossil fuels which may be discharged into the atmosphere from power stations or domestic fireplaces.

Geothermal energy. Energy derived from the heat of rocks at many hundreds of metres below the surface, by pumping water down to that depth and back to the surface again.

Germinal tissues. Those of the ovary or testis, in which the germ cells – the ova and sperm – develop.

Gigawatt of electricity (GW(e)). A rate of output or use equal to 1000 million watts. A gigawatt year of electricity corresponds to this output for a year.

Gonads. The ovaries or testes.

Graft, bone marrow. Intravenous injection of bone marrow cells from a donor in the hope of obtaining their multiplication and function in the marrow of an irradiated or other recipient.

GW(e)y. *See* Gigawatt.

Hard rock mines. Uranium or other metal mines, as distinct from coal mines.

Heavy water. Water containing the 'heavy' isotope of hydrogen (deuterium or hydrogen-2) in its structure.

Helix of DNA. The spirally wound double strand of DNA present in each chromosome.

Histone. A form of protein incorporated in the chromosome, having a structural rather than a genetic function.

ICRP. The International Commission on Radiological Protection (pp. 15, 163).

Image intensification. An electronic technique for increasing the brightness of x-ray images on fluoroscopic screens.

Incidence of a disease. The rate with which new cases of the disease occur, e.g. per year and per million population.

Induced radioactivity. Radioactivity caused by the effect e.g. of neutrons in converting stable atoms into radioactive ones.

Induction of cancer or genetic defects. Causation of these effects.

Introns. Sections of chromosomes found, or believed, not to code for the formation of protein molecules.

Ion. An electrically charged particle formed when one or more electrons are stripped off an atom. Ionization: the formation of ions. Ionizing radiation: types of radiation, with which the present text is concerned, which form ions when passing through body tissues or other materials.

Kiloton. Expresses the explosive yield of an atomic bomb as equal to that of 1000 tons of a conventional explosive (trinitrotoluene).

Kilowatt of electricity (kW). A rate of output or use equal to 1000 watts.

Labelling of molecules. By attachment of radioactive atoms onto chemical substances so that the presence of small quantities of the substances can be detected and measured.

Latent period, or latency, of cancer development. The time interval (years or decades) between the cellular change ('cell transformation') causing a potentiality for cancer formation, and the development of a detectable cancer (p. 126).

LET. Linear energy transfer, indicating the amount of the initial energy of a particle given up per unit length of its path through material, as a result of the ionizations and other effects caused. High LET radiations include those of alpha particles and neutrons. Beta and gamma radiations and x rays are of low LET.

Leukaemias. Forms of cancer due to multiplication of cells of bone marrow, with different types of leukaemia derived from different types of marrow cell, not all apparently inducible by radiation. In this text leukaemia is included in the term 'cancer', although distinguished from most other forms of malignant disease by not forming a 'solid cancer'.

Lymphocyte. A form of white blood cell.

Magnox. Type of reactor, operating in the UK since 1956, in which fuel rods are sheathed in magnesium oxide.

Malignancy, malignant disease. Terms (ordinarily) referring to cancer.

Man-sievert (man-Sv). Unit of collective dose, indicative of numbers of people exposed, and doses at which they are so exposed (p. 47).

Mantle of earth. Layers of rock extending from below the crust (q.v.) to a depth of about 3000 km (below which the remaining 3500 km is termed the core).

Megabyte. One million units of information, each 'unit' consisting of 8 simple (yes/no) alternatives.

Metabolite. Material formed in the course of, or as a result of, chemical processes (metabolism) in the body.

Microsievert. One millionth of a sievert.

Millisievert (mSv). One thousandth of a sievert.

Moderator. Material (e.g. graphite, heavy water) lying between fuel rods in the core of a reactor, to slow (i.e. reduce the energy of) neutrons released in fission, to increase their likelihood of causing further nuclear fissions.

Mortality from a disease. Rate at which deaths from the disease occur, e.g. per year and per million population.

mSv. *See* millisievert.

Muon. Unstable subatomic particle of mass intermediate between those of electron and proton.

Nanometre. Measure of length equal to one millionth of a millimetre.

Neoplasm. *See* cancer.

Non-stochastic effects. Those for which the severity of the effect varies with the size of the dose causing it, and for which the likelihood of the effect is high once a threshold dose is exceeded (as in Chapter 7).

Nuclear radiation. Radiation emitted from atomic nuclei, e.g. during their radioactive decay or fission, or when they are bombarded with electrons as in the production of x rays.

Nucleotide. The unit component in the structure of DNA which is repeated every few nanometres along the length of the DNA molecule. Each nucleotide consists of one of the four bases (q.v.), a sugar molecule (desoxyribose) to which it is attached, and a phosphate (phosphoryl) group through which each nucleotide is attached to the adjacent one in the chain (Fig. 8.1).

Oocyte. Progenitor cell from which the ovum is developed.

Organ. The meanings of organ and tissue overlap. 'Organ' is appropriate when all body cells of a certain type occur together in the same structure, e.g. in liver, lung, or brain. 'Tissue' applies when cells of a certain type occur in various parts of the body, e.g. those of bone marrow, connective tissue, or lymph glands. Thus, correctly, one must state that 'radiation affects cells of most of the organs and tissues' of the body.

Osteosarcoma. Cancer of bone cells, and predominantly of the endosteal cells (q.v.).

Phantom. Model e.g. of the human body constructed of materials allowing the same transmission and absorbtion of radiations as in the body, so that measurements can be made to estimate the dose received at different positions within the body.

Phosphoryl groups. The chemical groups (of formula = PO_4) linking successive sugar molecules along the 'backbone' of the DNA molecule.

Photometer. Device for measuring luminous intensity of light.

Pion, or pi-meson. Unstable subatomic particle of mass intermediate between those of electron and proton.

Platelets. Non-cellular structures present in blood, with the function of preventing escape of blood from damaged blood vessels. Formed by cells in the bone marrow.

Plating out, e.g. of iodine. Selective deposition and adherence of volatile materials present in air, onto the surface of a surrounding structure in which the air is contained.

Radionuclide. Any type of atom of which the nuclei are unstable and undergo radioactive decay. Distinct from radioisotope in that the latter refers to radionuclides of the same chemical element; thus carbon-14 is a radionuclide, and is a radioisotope of carbon.

RBE. Relative biological effectiveness; thus, for example, the RBE found for neutrons in a particular investigation is equal to the ratio between the absorbed dose (gray) of x rays required to cause a certain frequency of effects and the absorbed dose (gray) of neutrons causing the same frequency of effects.

Reprocessing of nuclear fuel. Chemical treatment of the contents of used fuel rods to extract unused fissile material including plutonium.

Ribosomes. Structures within the cytoplasm of cells in which proteins are formed.

RNA. Ribonucleic acid. In mammalian cells, certain forms of RNA transfer information from chromosomal DNA to ribosomes. The molecular structure is as in DNA except in the type of sugar (ribose not desoxyribose) and of one of the bases (uracil not thymine).

Scanning. Measuring the distribution of radioactivity over an area or length (linear scanning) of the body, by 'counting' the rate at which radiation is received (count rate) by an instrument (counter) in different positions over the body surface.

SI. The 'Système International' of units recommended in 1960, of which the radiation units now include the gray (p. 46), the sievert (p. 46), the becquerel (p. 68), and their submultiples.

Sievert (Sv). Unit of dose equivalent (p. 46).

Singleton. Child born in single, rather than twin, birth.

Somatic effects. Those expressed in body (somatic) organs or tissues, as distinct from genetic effects expressed in later generations, and due to action on germinal tissues (or 'gonads').

Sperm or spermatozoa. Mature male germ cells developed from spermatogonia and other earlier forms.

Static eliminator. Device for preventing the accumulation of static electricity, by causing ionization of surrounding air.

Stem cells. Groups of constantly dividing cells in various tissues which maintain a supply of 'adult', and sometimes non-dividing, cells in the tissue or in the blood.

Stochastic effects. Those for which the probability of the effect occurring, rather than the severity of the effect if it occurs, varies with the size of the dose. For such effects, as in the induction of genetic defects or cancer, no threshold dose can be assumed below which some effect may not occur.

Stratosphere. The atmosphere above a height of 10 to 16 km, this height varying with latitude.

Sucrose gradient analysis. Determination of the size of its molecules by the rate at which a material sediments in solutions containing different concentrations of sucrose.

Supernovas. Stars of which the outer layers are discharged explosively when the inner layers collapse inwards to form neutron stars (or black holes).

Sv. *See* sievert.

Tailings. Accumulations of rock or residues of ore from mines or milling operations, and from which uranium or other materials have been extracted.

Thymus. Gland in the upper part of the chest, concerned with maintenance of immune defences.

Tissue. *See* organ.

Titanates. Compounds containing titanium, found in various stable minerals.

TLD. Thermoluminescent dosimeters (p. 52).

Track. The route through tissue, etc., of a particle of any form of ionizing radiation, shown on photographic plates by a sequence of ionizations along or near that route.

Transformation, of cell. Change giving that cell the potentiality to multiply in an uncontrolled way and form of a cancer.

Triplet. Series of three consecutive bases in the DNA molecule constituting one 'word' in the genetic code.

Tritium. The radioactive isotope of hydrogen having mass number 3.

Troposphere. Atmosphere below a height of 10 to 16 km, this height (of the 'tropopause') varying with latitude, and being greatest at the equator.

Tumour. *See* cancer.

Ultrasonic measurement. Determination of the thickness or position of body structures by the way in which their boundaries reflect (ultrasonic) vibrations.

UNSCEAR. The United Nations Scientific Committee on the Effects of Atomic Radiation; set up by the General Assembly in 1955 and reporting to it.

INDEX

Pages shown in bold type contain definition or meaning of terms. Pages 187-93 refer to the Glossary.